MILITARY STRATEGY
IN TRANSITION

Also of Interest from Westview Press

Nuclear Deterrence in U.S.-Soviet Relations, Keith B. Payne

†*Thinking About National Security: Defense and Foreign Policy in a Dangerous World*, Harold Brown

The Half War: Planning U.S. Rapid Deployment Forces to Meet a Limited Contingency 1960–1983, Robert P. Haffa, Jr.

†*The Defense of the West: Strategic and European Security Issues Reappraised*, edited by Robert Kennedy and John M. Weinstein

†*Defense Facts of Life: The Plans/Reality Mismatch*, Franklin C. Spinney

†*U.S. Defense Planning: A Critique*, John M. Collins

Laser Weapons in Space: Policy and Doctrine, edited by Keith B. Payne

†*NATO—The Next Thirty Years: The Changing Political, Economic, and Military Setting*, edited by Kenneth A. Myers

Maneuver Warfare and the Marine Corps, Richard Scott Moore and William S. Lind

U.S. Military Power and Rapid Deployment Requirements in the 1980s, Sherwood S. Cordier

†Available in hardcover and paperback.

Studies in International Security Affairs and Military Strategy

Military Strategy in Transition:
Defense and Deterrence in the 1980s
edited by Keith A. Dunn
and William O. Staudenmaier

Current NATO military strategy is based on the policy of flexible response that U.S. and European politicians endorsed in 1967; for over 15 years, no fundamental changes in NATO's defense strategy have occurred. If NATO cannot stop a Warsaw Pact aggression conventionally, it continues to threaten a gradual and controlled nuclear escalation of both theater and strategic nuclear weapons.

Many analysts now question the fundamental principles underlying NATO's policy and strategy, given the enormous changes that have occurred in the strategic environment between 1967 and 1984. The contributors to this book examine the recent proposal by Samuel Huntington, who advocates that NATO adopt a conventional counter-retaliatory strategy based on offensive military actions deep into Eastern Europe. In evaluating this new proposal, the authors analyze the potential impact that it would have on U.S. and NATO military doctrine, assess probable European and Soviet reactions to NATO adopting a conventional counter-retaliatory strategy, and address the linkages existing between conventional and nuclear strategy. In the final chapter, the editors consider the policy, strategy, and force structure questions raised in the book and recommend policy options for the United States.

Keith A. Dunn is the senior policy analyst at the Strategic Studies Institute. **William O. Staudenmaier** is a colonel in the U.S. Army and director of strategy for the Center for Land Warfare, U.S. Army War College. They are coauthors of a forthcoming book, *Strategic Implications of the Continental-Maritime Strategy Debate.*

MILITARY STRATEGY IN TRANSITION: DEFENSE AND DETERRENCE IN THE 1980s

Edited by

Keith A. Dunn
William O. Staudenmaier

Routledge
Taylor & Francis Group

LONDON AND NEW YORK

To
Betty and Terry

First published 1984 by Westview Press

Published 2018 by Routledge
52 Vanderbilt Avenue, New York, NY 10017
2 Park Square, Milton Park, Abingdon, Oxon OX14 4RN

Routledge is an imprint of the Taylor & Francis Group, an informa business

Library of Congress Catalog Card Number: 84-51615

ISBN 13: 978-0-367-01652-4 (hbk)
ISBN 13: 978-0-367-16639-7 (pbk)

CONTENTS

PREFACE

This volume evolved from a conference held in July 1983 at the US Army War College. The conference participants drawn from both the academic world and government were brought together to discuss a new concept for NATO that Samuel P. Huntington has proposed: namely, NATO should adopt a conventional retaliatory offensive military strategy.

In this volume, thirteen specialists on US strategy and military doctrine, the Soviet Union, and Europe analyze the conventional retaliatory proposal with particular attention given to the effects such a proposal would have upon US military strategy and alliance relations. During the conference four major themes developed and in their own way each author dealt with one or more of these themes:

- What is the nature of deterrence?
- How much emphasis should be put upon offensive doctrines, tactics, operational strategies, etc., when evaluating NATO military strategy?
- If NATO changes its military strategy, how will this affect not only the Alliance, but also the Warsaw Pact?
- Has the strategic environment changed so significantly that major alterations in NATO's strategy are required?

A summary of the contributors' major points are found in Chapter 1. In the concluding chapter we concentrate specifically upon the US policy, strategy, and force structure implications associated with adopting a NATO conventional retaliatory strategy. By way of preview, it is fair to say, we think, that very few of the contributors to this volume completely endorse the Huntington proposal. Nevertheless, they recognize that analyzing and discussing new alternatives for NATO is a healthy endeavor given the significant changes which have occurred in the strategic environment since NATO was formed in 1949.

Our grateful appreciation is extended to LTG Richard Lawrence, who was Commandant of the US Army War College when this conference was held. The idea of the conference was his and without his personal support it would not have been a success. We gratefully acknowledge the generous support provided by the US Army War College Foundation, Inc., and its Executive Director, Colonel LeRoy Strong, USA (Ret.). Special thanks are due to

Deloris Hutchinson whose faithful efforts led to a completed manuscript. Also, we want to express our appreciation to the conference participants, who took time from their busy schedules to play an active role in the conference. Their comments helped the contributors sharpen their arguments in their final drafts. The conference participants are listed in the appendix. Working with the contributors has been a professional delight, and we would like to express our appreciation for their cooperation.

Neither this volume nor the individual articles in it should be construed to reflect the official position of the US Army War College, the Strategic Studies Institute, the Department of the Army, or the Department of Defense, unless so designated by other official documentation. The authors alone are responsible for any errors of fact or judgment.

Keith A. Dunn
William O. Staudenmaier
Carlisle Barracks, PA

CHAPTER 1

NEW STRATEGIES, NEW ALTERNATIVES: SOME INTRODUCTORY OBSERVATIONS

by

Keith A. Dunn and William O. Staudenmaier

In recent years, US defense scholars of widely varying political persuasions have begun to question the basis of contemporary US military strategy. Among many observers, there is a growing belief that the strategic environment has changed dramatically and, as a result, traditional US policies and military concepts must be adjusted to meet the new realities of this dynamic environment, particularly with respect to the strategic consequences of superpower nuclear parity. This has led the strategic community to search for ways to break the nuclear stalemate by reexamining conventional strategies and by ending the long-term neglect of military operational issues. The alternative proposals range from major reductions in ground and air forces stationed in Europe to greatly expanded naval and maritime forces for Third World power projection missions. Some analysts favor a reduction in foreign policy commitments as a way to balance strategic requirements with defense resources. In philosophical terms, the alternatives range from adopting a strategic defensive approach where the United States attempts to reduce or limit its political and military commitments to favoring a global strategic offensive in which the United States is prepared to engage not only the USSR, but other potential adversaries simultaneously in different theaters of operations.

This continuing strategic debate is hardly an academic exercise. There are important political as well as military implications associated with the idea that the United States needs to reexamine and possibly change the fundamental way that it has done its strategic business since the end of World War II. Clearly, any change in the US philosophy toward its basic strategic concepts should be reflected in changes to US military force structure and deployments. Also, any major changes to the basic strategic concepts would have to consider potential Soviet reactions, as well as the perceptions of major US allies. A strategy that would create dissension between the United States and its principal allies would hardly seem credible to the Soviets and, consequently, it would undermine rather than strengthen deterrence.

Samuel P. Huntington has offered an interesting and thought provoking alternative for US national security policy during the 1980s and 1990s. In a wide ranging article, Huntington argues that improvements to US military forces are required. But even more importantly—and the primary reason for this book—Huntington argues that "new strategic ideas and an entirely new emphasis on strategy and strategic thinking" are required if the United States is to deal successfully with the changing strategic environment.[1]

CONVENTIONAL DETERRENCE

Huntington's thoughts have sparked a significant debate within the international academic and policy community. Because of the importance of the issues raised, as well as heightened interest within some circles of the Pentagon, the US Army Strategic Studies Institute organized a conference that centered on Huntington's belief that the United States must adopt a conventional retaliatory

2

strategy for Europe. The contributors to this book were invited to prepare papers that focused on major issues associated with Huntington's concept and present them at the symposium. In some instances, the authors supported portions of Huntington's thesis, but, in other cases, they strongly disagreed with his suggestions and proposals.

While it would not be accurate to say that the participants reached a consensus during the symposium, the following major themes or issues repeatedly emerged and weaved their way throughout the conference proceedings.

• What is the nature of deterrence? Has the US nuclear deterrent which is based upon a graduated controlled nuclear escalation response to a Soviet attack in Europe become discredited given the loss of US strategic nuclear superiority? To what extent can improvements to conventional capabilities strengthen deterrence?

• In US military strategy, what is the proper mix and emphasis on offensive doctrine, tactics, operational strategies, capabilities, etc., compared with defensive strategies?

• How would US allies and the Soviet Union and its allies perceive a proposal that envisioned an Allied conventional invasion of Eastern Europe? Would their reactions make it more difficult or easier to achieve declared interests and objectives?

• Has the strategic environment really changed to the extent that Huntington and others have argued?

At least one or more of these themes are addressed in each of the following chapters.

Can the United States effectively use conventional forces for deterrence by putting at risk something that the USSR highly values? This is Huntington's central point in his design of a "conventional retaliatory strategy" to enhance deterrence in NATO. In Chapter 2, Huntington elaborates upon this new concept in an effort to demonstrate why he believes conventional retaliation is "strategically desirable and militarily feasible." He presents a combination of reasons to justify his conventional retaliation strategy. They include a belief that:

• A consensus within Europe is developing which is more favorably inclined toward enhancing NATO's conventional rather than nuclear deterrence capabilities.

• A totally defensive military strategy is not credible. For a deterrent strategy to be credible, it must threaten retaliation at an aggressor's valued assets. Retaliation has always been a part of nuclear deterrence strategy and there is no reason why it should not be part of conventional strategy.

• A conventional retaliatory strategy is fully compatible with flexible response and forward defense because such a strategy should make flexible response more flexible and forward defense more forward.

• A "prompt allied offensive into Eastern Europe would stimulate . . . disaffection [within the Warsaw Pact] at the very start of the conflict."

• The threat of "a prompt allied offensive into Eastern Europe" would enhance deterrence in two ways. First, it would cause Moscow to stop thinking only in offensive terms and force the Soviets to reallocate forces for defense within Eastern Europe. By weakening Soviet offensive potential, Warsaw Pact chances of a successful invasion of Western Europe would be reduced thereby enhancing deterrence. As Huntington argues in Chapter 2, "a counteroffensive threat, in short, will lead the Soviets to strengthen their offense; a retaliatory offensive threat will lead them to weaken it."[2] Second, it would undermine the Warsaw Pact's resolve to support Moscow because it would be clear to East Europeans that NATO would not allow a conflict to occur where NATO members would suffer the brunt of conventional battle, while East European nations suffered few adverse consequences. NATO would not simply withdraw across Europe, trade space for time, stop the invasion, and then fight back to establish the status quo ante. Eastern Europe would suffer significant damage also as NATO forces execute a conventional retaliatory strategy.

• A conventional retaliatory strategy will contribute to deterrence not only in Europe, but also in other theaters. A NATO strategy for offensive military actions in Eastern Europe—an area highly valued by the USSR—would cause Moscow to think twice before initiating military actions in Southwest Asia or China. According to Huntington, linking events that occur in one theater to actions that occur in another theater will enhance deterrence: "If deterrence is to be reasonably well assured . . . , it must rest on the high probability that Soviet military action in any one area will also involve the Soviet Union in military hostilities in other areas." In other words, "horizontal/geographic escalation" is a good idea because it allows the United States to threaten "Soviet vital interests in an area other than the one which the Soviets are threatening."

OPERATIONAL CONCEPTS AND DOCTRINES

The primary focus of the "conventional retaliatory strategy" is deterrence. While Huntington elaborates upon his concept and briefly discusses ways to execute the concept in Europe in this volume, he gives scant attention to the operational aspects of his strategy or the forces required to carry out a retaliatory offensive. Therefore, when the conference was organized, it was clear that some discussion concerning how American and European military planners believed battles in Europe should be fought was necessary to determine if these concepts were compatible with the retaliatory approach.

Chapters 3 and 4 focus on two different operational doctrines for European defense: the US Army's approach (AirLand Battle) and the Supreme Allied Commander Europe's (SACEUR) concept (Follow-On Force Attack). Both doctrines embody deep attack concepts—i.e., striking at the Warsaw Pact rear echelon forces to disrupt and delay these forces—and as such to some degree support a conventional retaliatory offensive strategy. However, once one delves below the surface commonality, several major differences between AirLand Battle and Follow-On Force Attack emerge. First, AirLand Battle is primarily concerned with forces in the forward battle area, particularly the Soviet first echelon forces and their immediate reinforcements that face US Corps. SACEUR's concept of Follow-On Force Attack, on the other hand, is less concerned with attacking units than it is with interdicting fixed targets such as key transshipment points or other installations that affect Soviet capabilities to bring its strategic reserve forces to bear. AirLand Battle is "designed to win battles, not wars," according to Richard Sinnreich. On the other hand, the authors of Chapter 4 argue that Follow-On Force Attack primarily is concerned with deterrence and war limitation if deterrence fails. Second, both concepts require significant tactical air assets to be successful, but the requirements of each doctrine are not completely compatible. If AirLand Battle is to be successful, Corps commanders must receive battlefield interdiction (150-200 kilometers) support from the Air Force. The SACEUR plan calls for more preplanned air strikes across the entire theater and centralization of a large proportion of air assets at theater level. AirLand Battle requires decentralization so air assets will be responsive to Corps commanders. Coordination of battlefield tactical air strikes and interdiction in accordance with AirLand Battle doctrine is designed to enable the Corps commander to maneuver effectively in order to defeat Soviet forces.[3] Third, different technologies and acquisition policies will be needed for each concept. To be successful, both AirLand Battle and Follow-On Force Attack require that military commanders have the capability to acquire and strike targets beyond the immediate battlefield. The primary difference between these two approaches, however, is the depth to which the Corps commander needs to acquire and attack Soviet targets. AirLand Battle is dependent upon acquiring targets up to 150-200 kilometers from the forward edge of the battle area (FEBA), while SACEUR's concept necessitates acquiring and striking targets much deeper into Warsaw Pact territory (see Figure 4-1).

Who will provide the money to support such concepts? SACEUR believes that NATO can pay for Follow-On Force Attack if member states increase their defense spending four percent in real terms over the next five years.[4] Most NATO states have not met the three percent pledge that they made in 1978. Therefore, there is probably little reason to believe that they will meet a higher standard when domestic economic problems continue to plague Europe. If this is so, choices will be required in the material acquisitions policy because it is unlikely that the United States or NATO will be able to fund both AirLand Battle and Follow-On Force Attack simultaneously, if European defense budgets do not increase significantly. In the United States the financial issue is even more complicated because the two deep attack options for European defense are not the only new strategic concepts competing for Pentagon defense budget funds. For example, Secretary of the Navy John Lehman is striving to build a fifteen carrier Navy that will have the capability to fight in various theaters simultaneously. This in its own right is an expensive venture. In current dollars, it will cost between nine to ten billion dollars to increase the number of carriers from twelve to fifteen. To buy the carriers, air wings, and associated support ships will cost more than thirty-five billion dollars and this does not account for the lifetime operating costs for the new carrier battle groups.

The editors firmly believe that strategic debates should center on interests, objectives, and threats and not have budget issues as the central focus. Therefore, the Sutton, *et. al.,* caution that "what must be avoided at this stage are procedures or procurement that lock us into either approach [AirLand Battle or Follow-On Force Attack]" is important. Nevertheless, a credible and realistic military strategy must be financially acceptable. A failure to consider the financial implications of suggested military strategies or doctrines could result in the creation of a strategy which the American public will not financially support because of its expense.

EUROPEAN AND SOVIET RESPONSES TO CONVENTIONAL DETERRENCE

Perceptions—how other nations will view US actions and tailor responses to them—are also an integral part of any strategic analysis. Much of the conference's discussion centered around such perceptions. Catherine Kelleher analyzes European attitudes and possible reactions to a US conventional retaliatory strategy. Vernon Aspaturian discusses Soviet perceptions in a broad perspective, as well as examining the vulnerabilities and strengths of the Soviet inner and outer empires. Daniel Papp wrestles with the question: "How would the USSR respond if NATO adopted a conventional retaliatory offensive?" He presents nine plausible political-diplomatic and military responses that the Kremlin could choose to counter NATO's actions.

In Chapter 5, Kelleher disagrees with Huntington's suggestion that a growing consensus in Europe favors conventional defense. Despite increased European interest in conventional defense issues, traditional European concerns still exist about severing the linkage to US strategic nuclear weapons and making Europe the conventional battleground while the superpowers avoid damage from a European war. Europeans conclude that a forward defense is still the best NATO strategy because Europe lacks sufficient territory for a defense in depth. As Kelleher points out, some defense analysts in the Federal Republic of Germany are now more willing to stress conventional defense than in the past. Such attitudes, however, are not reflected in public defense debates in the United Kingdom or the Benelux states.

Domestic political considerations, Kelleher argues, probably will increase European reluctance to accept conventional defense strategies. Few European leaders are optimistic about economic growth over the next decade. This pessimism directly affects their attitudes toward improvements to conventional defense which analysts on both sides of the Atlantic agree would be expensive. Moreover, there is widespread belief in Europe that the United States is more interested with confronting the USSR rather than trying to negotiate, deal, and limit the tensions between the superpowers. As a result, some Europeans see ideas for enhanced NATO conventional defense as a way for the United States to get Europe to provide foot soldiers for NATO's defense so that the United States can confront Moscow in other theaters, particularly Southwest Asia. This is not an appealing idea to Europeans for at least two reasons. First, Europeans obviously believe that NATO is and should be the most important theater. While realizing that events outside of Europe affect NATO's security, anything that draws US attention too far away from Europe is by definition bad. Second, Europeans want to insulate NATO from the global US-Soviet competition. The United States may believe that detente was a failure, but Europeans generally regarded it as a success.

In Chapter 6, Aspaturian looks at Huntington's assumptions from a East European perspective and concludes that there would be no widespread support for a NATO retaliatory strategy within the region. Also, he questions the assumption that such a strategy would automatically undermine East European reliability. Exactly the opposite could result. East Europeans—particularly East Germans, Czechs, and Poles—are not enamored with the prospect of a reunified Germany under West German auspices. The memories of World War II and Hitler run just as deep in Eastern Europe as they do in the Soviet Union. According to Aspaturian, East European reliability "will depend almost entirely upon how they perceive the outcome." If NATO is believed to be winning, East Europeans will provide as little support for the USSR as is feasible. However, if the USSR appears to have the upper hand, East Europeans will want to be on the winning side.

Aspaturian also thinks that Huntington's notion of avoiding the nuclear weapons dilemma by adopting a conventional retaliatory strategy is flawed. East European security is vital to the defense of the Soviet inner empire—despite the development of long range bombers and theater nuclear weapons that are capable of attacking the USSR from bases in NATO, and despite the development of intercontinental ballistic missiles (ICBMs). Therefore, Aspaturian argues that the Soviet Union will use every means available, to include the "resort to nuclear weapons to protect this Soviet survival interest." The linkage between conventional strategies and nuclear weapons, which Keith Payne also addresses in Chapter 8, is a critical issue that the conventional retaliatory thesis overlooks, but it is a necessary requirement when a military strategist considers new alternatives for defense and deterrence for the 1980s. A primary task of the military strategist is to insure that new alternative strategies do not needlessly exacerbate the likelihood of nuclear warfare. Aspaturian's admonishment to avoid the typical American tendency to look for single instruments to enhance deterrence (more emphasis upon nuclear weapons versus increased conventional forces) is an important reminder. Nuclear weapons exist; no amount of wishful thinking will cause them to go away. We need to remember this fact as we consider new alternatives to old military strategies.

Aspaturian concludes by suggesting that a better approach might be to adopt a European territorial defense strategy. This is not a new idea. However, like the conventional retaliatory strategy approach, it would require a fundamental change in NATO's political outlook. Inherent in any territorial defense concept is the idea of trading space for time and wearing out an adversary as it confronts a prolonged partisan war. This is an anathema to West German politicians. While we do not necessarily disagree with Aspaturian's claim that a territorial defense approach presenting the USSR with a "vision of a protracted nonnuclear conventional and partisan war . . . is more likely to deter the Soviet Union than any other nonnuclear force, save the perception of an overwhelmingly superior NATO conventional force," we find that this approach is as politically questionable as a conventional retaliatory offensive strategy. To make either strategy possible, NATO would have to change the policy guidelines under which current military strategists must operate.

In Chapter 7, Papp concentrates on nine political-diplomatic and military responses that the Soviets could adopt if NATO pursued a conventional retaliatory offensive strategy. These responses range from increased propaganda activity against the idea in Western and Eastern Europe and the Third World to more advanced Soviet deployments in the forward areas in Eastern Europe. Generally, Papp is no more sanguine than Aspaturian that the new strategic alternative would be welcomed automatically by East Europeans. As he says, "the announcement of a NATO conventional retaliatory strategy may offer some hope [to those East European states already inclined to oppose the USSR] and through hope a willingness to oppose Soviet policies and perspectives." However, it equally could "enhance the credibility of Soviet claims of Western expansionism and hostility" thereby solidifying Soviet control in the region.

Also, Papp argues that the adoption of a new NATO strategy could contribute to crisis instability. If the retaliatory force had to be maintained at a high stage of readiness to execute this operational strategy (as it would) and Moscow believed that NATO was intent upon carrying out such a strategy, this could push the USSR even further toward preemption. Before adopting a conventional retaliatory offensive strategy, Papp warns that US decisionmakers must be confident that such a move would not destroy NATO's peacetime political consensus and that the additional forces to initiate such a strategy can be obtained. If these two conditions cannot be met with a reasonable level of confidence, the proposal should be discarded as impractical, because it would create more risks than potential benefits.

NUCLEAR DETERRENCE LINKAGES

In Chapter 8, Keith Payne draws our attention to the realities of the current nuclear era. How does the United States cope with nuclear weapons as it develops a comprehensive strategy to achieve its interests and objectives? Payne does not argue directly for or against the conventional retaliatory proposal. He does, however, make the point that strategists—particularly when considering nuclear options—must consider more than offensive approaches. Strategic nuclear defensive capabilities, he claims, are also essential for deterrence.

Huntington believes that NATO's conventional strategy has overemphasized deterrence by denial rather than deterrence by punishment which he favors. As a result, his conventional retaliatory offensive strategy is an attempt to rectify this imbalance by providing the United States the conventional capabilities to threaten something that is of value to the USSR and to punish Moscow with conventional means if deterrence should fail in Europe. Deterrence through punishment has always been a part of US nuclear strategy and Huntington believes it could work conventionally also.

Payne, on the other hand, supports a "balanced deterrent" and nuclear force posture because this is "the only school of thought that makes provision for the possible failure of deterrence." The balanced deterrent approach, according to Payne, would emphasize not only modernization of offensive nuclear weapons, but also improved Ballistic Missile Defense, civil defense, and strategic air defense systems to insure survivability for the National Command Authority, and C^3I (command, control, communications, and intelligence) facilities. In essence, Payne argues that the United States needs to enhance its deterrence through denial capabilities for its strategic nuclear forces, while Huntington favors deterrence through punishment for conventional forces.

Together Payne and Huntington force one to think about the linkages between conventional and nuclear deterrence. Currently, it seems to be fashionable to believe that one can set aside nuclear questions and focus upon conventional deterrence.[5] In reality neither can be considered in isolation. Military strategists must consider how alternatives which are primarily conventional in focus can have implications for nuclear escalation. In the era of nuclear parity, it is unwise to attempt to compensate for conventional force structure deficiencies by advocating operational strategies that could increase the risk of nuclear escalation. Moreover, the military strategist must look beyond the stated purpose of a strategy and analyze the implications associated with it. One does not have to accept Payne's conclusion or force structure recommendations, but the military strategist cannot escape his charge to think beyond

11

deterrence. Implicit in that charge also is the requirement to think about how conventional warfighting options relate to nuclear deterrence, warfighting, and damage limitation.

EUROPEAN DETERRENCE

In the final chapter, the editors return to the fundamental issue raised by Samuel Huntington. Namely, to restore deterrence in Europe, the Soviet Union must be confronted with a threat of a conventional retaliatory offensive aimed at the heart of their inner empire—Eastern Europe. Building on the analyses provided by the other authors, the concluding chapter seeks to put the debate in perspective regarding the four major themes of the conference and to answer some of the operational military questions raised, but never satisfactorily answered by the conventional retaliatory offensive thesis.

Chapter 9 begins by noting that most military analysts agree that the strategic environment has indeed changed, since the assumptions on which US strategy is based were formulated over thirty years ago. But has that environment changed so dramatically that the basic US strategic assumptions have been invalidated? Here we argue the evidence is mixed, particularly with respect to the central strategic issue between the superpowers—conflict avoidance. Given a strategic environment in which neither the Soviet Union nor the United States seem likely to achieve a first strike nuclear capability, it appears that the unarticulated superpower policy of avoiding situations in which a direct military confrontation appears likely should continue.

Next, the editors focus on the viability of deterrence in an era of nuclear parity. Strategists, who have been grappling with the problems raised by the condition of strategic nuclear parity, have sought new ways to deter superpower conflict. However, two new proposals—horizontal escalation and conventional offensive land strategies—beg the following questions:

• How well do these new concepts achieve US interests and objectives?

• What impact do these concepts have on escalation dynamics?

Three generic issues raised by horizontal escalation or war-widening strategies, whether by land or sea, are examined. Questions of escalation, proportionality, and war termination must be answered. First, there are dangers that war-widening strategies will lead to escalations of the political objective by raising the military ante resulting in vertical escalation and nuclear war. Second, what Soviet objectives or possessions outside Europe, Southwest Asia, and Korea/Japan could be threatened that would provide a proportional return to the West for the loss of its critical areas? Finally, it is relatively easy to begin horizontal escalation, but extremely difficult to predict where it will end.

Chapter 9 concludes by considering the military consequences associated with the conventional retaliatory offensive proposal and some of the other war-widening concepts. The editors discuss the significant force structure implications of a conventional retaliatory offensive into Eastern Europe. Other military and political problems that concern stationing of military forces, US political commitments required in peacetime in advance of a crisis, and the financial and social costs that are associated with such proposals also are analyzed.

In recent years, two ideas have dominated the study and practice of military strategy. One is the maturing reality of superpower nuclear parity. This has convinced many defense analysts that the only value of nuclear weapons is to deter the use of other nuclear weapons. This perception has led to the second notion—only conventional strategies can break the nuclear stalemate. By posing a conventional deterrent concept that cuts to the heart of NATO's dilemma, Huntington has forced the national security affairs community to focus on basic concepts and issues which all strategists must deal with in the coming years. In coming to grips with these issues, choices will have to be made on some of the most perplexing strategic issues that have faced the West since the end of World War II. Choices require knowledge and knowledge should improve strategy. The authors of this volume have made a major contribution in expanding our knowledge of the subtleties of conventional deterrence. It remains to be seen if this knowledge can be translated into effective policy and strategy.

ENDNOTES

1. Samuel P. Huntington, "The Renewal of Strategy," in *The Strategic Imperative: New Policies for American Security,* ed. Samuel P. Huntington (Cambridge, Mass.: Ballinger Publishing Company, 1982), p. 50.

2. The distinction that Huntington makes between counteroffensive and retaliatory offensive is important. A counteroffensive is one which begins *after* the Soviet invasion of NATO territory has been stopped. A retaliatory offensive is one that begins *simultaneously* with the first Soviet unit entering NATO territory.

3. The allocation procedures for close air support for the two concepts are also incompatible. AirLand Battle requires that allocation of preplanned air sorties be decentralized to the Corps level and that the ground commander designate target priority. Follow-On Force Attack, however, demands that the allocation be made at theater level, with target priorities designated by the SACEUR.

4. Bernard W. Rogers, "The Atlantic Alliance: Prescriptions for a Difficult Decade," *Foreign Affairs,* vol. 60, no. 5 (Summer 1982), pp. 1145-1156.

5. John J. Mearsheimer, *Conventional Deterrence* (Ithaca, NY: Cornell University Press, 1983).

CHAPTER 2

CONVENTIONAL DETERRENCE AND CONVENTIONAL RETALIATION IN EUROPE

by

Samuel P. Huntington*

For a quarter century the slow but continuing trend in NATO strategy—and in thinking about NATO strategy—has been from emphasis on nuclear deterrence to emphasis on conventional deterrence. When it became clear that the famous Lisbon force goals of 1952, embodied in MC 14/1, had no hope of realization, NATO strategy appropriately stressed the deterrent role of nuclear

*This chapter elaborates in a more refined and detailed form arguments which I originally set forth in "The Renewal of Strategy," in *The Strategic Imperative: New Policies for American Security,* ed. Huntington (Cambridge, Mass.: Ballinger, 1982), pp. 21-32, and in "Broadening the Strategic Focus," in *Defense and Consensus: The Domestic Aspects of Western Security, Part III, Adelphi Papers,* No. 184 (London: International Institute for Strategic Studies, 1983), pp. 27-32. In a few spots in this essay I have shamelessly plagiarized these earlier writings. I am grateful to Richard K. Betts and Eliot Cohen for their helpful critical comments. This chapter was originally prepared for the US Army War College Military Policy Symposium— "Defense and Deterrence in the 1980s: New Realities, New Strategies." Subsequently, it appeared in *International Security* and is reprinted from that journal with permission of the MIT Press. Copyright 1984 by the President and Fellows of Harvard College and the Massachusetts Institute of Technology.

weapons, in terms of both massive retaliation by US strategic forces and the early use of tactical nuclear weapons in Western Europe. This strategy was codified in MC 14/2 in 1957. Shortly thereafter, however, the development of Soviet strategic nuclear capabilities and, more particularly, the massive deployment by the Soviets of theater nuclear weapons raised serious questions as to the desirability of NATO's relying overwhelmingly on early use of nuclear weapons to deter Soviet attack. In the following years, the emphasis shifted to the need for stronger conventional forces capable of mounting a forward defense of Germany for a period of time and to a strategy of flexible response, in which, if deterrence failed and if conventional defenses did not hold, NATO would have the options of resorting to tactical, theater, and eventually strategic nuclear weapons. In 1967 this strategy became official NATO policy in MC 14/3.

The past several years have seen increasing support for shifting the deterrent emphasis even further in the conventional direction. This perceived need derives, of course, from the facts of strategic parity between the United States and the Soviet Union, Soviet achievement of substantial predominance in theater nuclear forces, and a continued and, in some respects, enhanced Soviet superiority in conventional forces. In these circumstances, in the event of a successful Soviet conventional advance into Western Europe, how credible would be the threat of a nuclear response? In the face of Soviet superiority at that level, why would NATO resort to theater nuclear weapons, with all the destruction to both sides that would entail? Even more significantly, why would the United States use or even threaten to use its strategic nuclear forces, if that would ensure massive Soviet retaliation against North America? The concerns which DeGaulle articulated (even if he may not have believed them) in the early 1960s had by the early 1980s come to be first believed and then articulated by a broad spectrum of statesmen and strategists. The standard reassurances of the validity of the American nuclear guarantee, as Henry Kissinger put it in 1979, "cannot be true" and "it is absurd to base the strategy of the West on the credibility of the threat of mutual suicide."[1] Even McGeorge Bundy, who immediately countered Kissinger's statement with an argument for the continued efficacy of nuclear deterrence in Europe, dramatically abandoned that position three years later.[2]

Current NATO strategy also has little support among Western publics. In 1981, in the four major Western European countries, for instance, overwhelming majorities (66 percent in Germany, 71 percent in Britain, 76 percent in France, 81 percent in Italy) favored either no NATO use of nuclear weapons "under any circumstances" or NATO use only if the Soviet Union "uses them first in attacking Western Europe." In all four countries only small minorities (12 percent in Italy, 17 percent in France and Germany, 19 percent in Britain) supported existing NATO strategy that "NATO should use nuclear weapons to defend itself if a Soviet attack by conventional forces threatened to overwhelm NATO forces."[3] Somewhat similarly, in the United States, public opinion generally opposed a "no first use" declaration but also by an overwhelming margin (62 percent) answered "no" to the question as to whether the United States "would *ever* be justified in using nuclear weapons first during a war against another country."[4] In democratic societies, expert opinion and public opinion often differ on nuclear weapons issues. In the West today, however, they agree in rejecting reliance on the use of nuclear weapons to respond to conventional attack. In its current formulation, flexible response is seen as inadequate by the strategists, unsupportable by the public, and, one must assume, increasingly incredible by the Soviets.

The conclusion almost universally drawn from this perceived deteriorating credibility of the nuclear deterrent to Soviet conventional attack in Western Europe is the need to strengthen NATO conventional forces. The desirability of doing this is broadly supported by conservative, liberal, and, in Europe, socialist politicians. It has been endorsed in one form or another by a wide variety of military experts and strategists, including General Bernard Rogers, Professor Michael Howard, Senator Sam Nunn, the Union of Concerned Scientists study group, the No First Use "Gang of Four," the American Academy of Arts and Sciences European Security Study, informed Social Democratic Party (SPD) analysts, the Reagan Administration, and, so far as one can gather, those Democratic presidential aspirants who have addressed the issue.[5] The conventional wisdom is, in short, that stronger conventional forces are needed to enhance conventional deterrence and thus compensate for the declining effectiveness of nuclear deterrence.

THE REQUIREMENTS OF CONVENTIONAL DETERRENCE

The conventional wisdom suffers from two significant weaknesses.

First, NATO countries are unlikely to commit the resources necessary to achieve the required strengthening of NATO defenses. Ever since the Lisbon conference, various efforts have been made to bolster NATO's conventional capabilities so as to decrease reliance on nuclear retaliation. Except for a brief period in the mid-1960s, these efforts have not been notably successful. After a quarter of a century, deterrence by conventional forces remains appealing, but it also remains an unreality. For understandable reasons, European governments and publics have been unwilling to appropriate the funds and make the sacrifices that would be required to make it effective. This pattern continues. In 1978 the Alliance committed itself to the Long Term Defense Program requiring 3 percent annual increases in defense spending by its member countries. Apart from the United States, however, the members of NATO have, with occasional exceptions, generally failed to meet that goal. General Rogers now argues that an effective conventional defense for NATO can be achieved if its members increase their military spending by 4 percent annually. The European Security Study comes to a similar conclusion. But if NATO countries have failed to achieve a sustained 3 percent increase, how realistic is it to talk of 4 percent increases? The attitudes of European publics and governments do not seem to be more favorable to voting larger defense budgets than they have been in the past, and economic conditions for such increases are at present far less propitious. This does not mean that no increases in NATO conventional defense capability will occur. Clearly they will. The budget increases are likely, however, to be no more than 2.0-2.5 percent, and hence they will not achieve the levels thought necessary by those who see increased conventional capability as the solution to the problem of deterrence. Thus, while nuclear deterrence of a Soviet conventional attack on Western Europe suffers from a lack of credibility, conventional deterrence of such an attack suffers from a lack of capability.

The second problem with the strengthening-conventional-forces approach is more serious. It concerns not inadequate resources but erroneous, if generally unarticulated, assumptions. It would still be present in some form even if NATO defense spending did increase by 4 percent a year. It is much more salient if that goal is not achieved. It involves the requirements of deterrence.

Military forces can contribute to deterrence in three ways. First, they may deter simply by being in place and thus increasing the uncertainties and potential costs to an aggressor, even though they could not mount an effective defense. Allied forces in Berlin have performed this role for years, and the argument for being able to move airborne forces and Marines rapidly to the Persian Gulf, in the event of a Soviet invasion of Iran, rests on a similar premise. Simply the presence of American forces in Khuzistan might deter the Soviets from moving in on the oil fields. Second, military forces can deter by raising the possibility of a successful defense and hence forcing the aggressor to risk defeat in his effort or to pay additional costs for success. This has been the traditional deterrent role assigned to NATO forces in Germany. Third, military forces can deter by threatening retaliation against assets highly valued by the potential aggressor. This, of course, has been the classic role of strategic nuclear forces. Unlike deterrence by presence or deterrence by defense, however, this form of deterrence is not effective simply because the requisite military capabilities exist; it requires a conscious choice by the defender to retaliate; and hence the aggressor has to calculate not only the defender's capabilities to implement a retaliatory threat but also the credibility of that threat.

One of the striking characteristics of the new conventional wisdom is the extent to which stronger conventional defenses are identified with a stronger conventional deterrent. If only NATO can enhance its military defenses, Soviet aggression will be deterred: this assumption is implicit in most of the arguments for stronger NATO forces and it is at the heart of the report by the European Security Study. That report, indeed, treats conventional defense and conventional deterrence as virtually interchangeable concepts. The title of the report is "Strengthening Conventional Deterrence in Europe"; the central sections of the report deal with "The Specific Requirements for Effective NATO Conventional Defense" and "The Means for Enhancing NATO's Conventional Defensive Capability"; throughout it is assumed that improved conventional defenses mean improved conventional deterrence.

To a limited degree, this assumption is, of course, justified. The stronger NATO forces are, the greater the investment the Soviets would have to make to achieve a given set of goals in Western Europe. Yet the easy identification of deterrence with defense flies in the face of logic and in the face of long-standing traditions in strategic thought. One of the landmark works on this subject (still valuable after twenty years), *Deterrence and Defense* by Glenn Snyder, is, indeed, based on the opposition between defense and deterrence and the extent to which strategies and forces appropriate to serve one goal may not be suited to achievement of the other. In addition, deterrence itself, that is, the effort to influence enemy intentions, may be pursued through both "denial capabilities— typically, conventional ground, sea, and tactical air forces" and "punishment capabilities—typically, strategic nuclear power for either massive or limited retaliation."[6] In the years since Snyder, strategists generally have tended to make the same identification. In the process, concern with the distinction between nuclear and conventional capabilities has tended to obscure the equally important distinction between defensive and retaliatory capabilities. In current NATO planning, nuclear and conventional capabilities can both be used for defensive purposes; only nuclear capabilities can be used for retaliatory purposes. Eliminating or drastically downgrading nuclear forces means eliminating or drastically downgrading the retaliatory component that has always been present in NATO strategy. Those who argue for conventional defense are, in effect, arguing for deterrence without retaliation. This is a fundamental change in NATO strategy, at least as significant in terms of deterrence as the shift from nuclear to conventional forces. For as both logic and experience make clear, a purely denial strategy inherently is a much weaker deterrent than one which combines both denial and retaliation.

For a prospective attacker, the major difference between denial and retaliation concerns the certainty and controllability of the costs he may incur. If faced simply with a denial deterrent, he can estimate how much effort he will have to make and what his probable losses will be in order to defeat the enemy forces and achieve his objective. He can then balance these costs against the gains he will achieve. He may choose zero costs and zero gains; he may decide to limit his gains to what can be achieved by a given level of costs; he may decide to incur whatever costs are necessary to achieve the gains he desires. The choice is his. If, however, he is confronted with a retaliatory deterrent, he may well be able to secure the gains he wants with relatively little effort, but he does not know the total costs he will have to pay, and those costs are in large measure beyond his control. The Soviet general staff can give the Politburo reasonably accurate estimates as to what forces it will require and what losses it will probably suffer to defeat NATO forces in Germany and extend Soviet control to the Rhine. For years, however, it could not predict with any assurance whether US nuclear retaliation to such a move would be directed to battlefield targets, military targets in Eastern Europe and/or the Soviet Union, or industrial and population centers in the Soviet Union. Precisely this uncertainty and absence of control made the threat of retaliation a strong deterrent. If these problems of uncertainty and uncontrollability are eliminated or greatly reduced, the effectiveness of the deterrent is seriously weakened.

The difficulties of relying on deterrence by defensive means have long been emphasized in the strategic field. No defense system—antiaircraft, ABM, or civil defense—deployable now or in the foreseeable future could prevent some nuclear weapons from reaching their targets and causing unprecedented destruction. Hence deterrence of an attack must depend upon the ability to retaliate after absorbing the attack. Much the same is true at the conventional level. In the past, conventional deterrence has usually meant deterrence-by-denial, and the frequency of wars in history suggests that this conventional-denial deterrence was not often effective. Nor has it been effective in the modern era. In a careful survey, John Mearsheimer identified twelve major instances of conventional deterrence between 1938 and 1979. In two of these cases, deterrence worked; in ten, deterrence failed.[7] This 83.3 percent failure rate for deterrence by conventional defense after 1938 contrasts rather markedly with the zero failure rate for deterrence by nuclear retaliation for a quarter century after 1945.

An initial offensive by a strong and determined attacker, particularly if accompanied by surprise, inevitably will score some gains. As Saadia Amiel summed up the lessons of the 1973 Arab-Israeli war and the implications of precision guided munitions (PGMs): "without very clear offensive options, a merely passive or responsive defensive strategy, which is based on firepower and fighting on friendly territory, cannot withstand an offensive strategy of an aggressor who possesses a relatively large, well-prepared standing offensive military force."[8] This is certainly the case in central Europe. Given NATO's current conventional defenses and any improvements in them, a Soviet conventional offensive in Europe is, inevitably, going to be at least a partial success. The Soviets may not reach the Pyrenees, or the English Channel, or even the Rhine. They may or may not occupy Frankfurt, Hamburg, or Munich. Inevitably they will, however, score some gains. They may pay a substantial price in losses of men and equipment, but they will still occupy West German territory

21

and conquer West German population and industry. That is a certainty produced by geography and any realistically conceivable balance of conventional forces in central Europe. Given the nature of existing forces and strategies, these gains could well be substantial, but that is not necessary for the argument.

Assume that the Soviet offensive does grind to a halt after Soviet forces have occupied a greater or lesser portion of West Germany. What then? In theory, the Allies should bring in their reinforcements from North America and put together a counteroffensive to drive the Soviets back. This would, however, be an extraordinarily difficult military and logistical undertaking. Inevitably the pressures would be on all parties to attempt to negotiate a cease-fire and a resolution of the conflict. With their armies ensconced in Hesse, Lower Saxony, and Bavaria and the differing interests of the Allies manifesting themselves, the Soviets would clearly have the upper hand in such negotiations. It takes little imagination to think of the types and appeals the Soviets would make to West German authorities and political groups to accept some degree of demilitarization or neutralization in order to secure Soviet withdrawal and to avoid the replay of World War II in their country.

A Soviet invasion of West Germany that ended with the neutralization and/or demilitarization of all or part of that country would be a tremendous success from the Soviet point of view. It would decisively alter the balance of power in Europe and in the world. Its costs, in terms of losses of men and equipment, would have to be very substantial to outweigh these political, military, and diplomatic gains. In 1939 and in 1941, once they had devised means to neutralize possible Allied retaliation by bomber attacks on cities, Hitler and the Japanese launched their offensives into Poland and southeast Asia expecting, not entirely unreasonably, that their democratic opponents would lack the staying power to deprive them of their initial territorial conquests.[9] In the absence of a credible retaliatory threat against valued Soviet assets, the Allies would be tempting fate to assume that the Soviets would not be tempted to make a comparable move into Western Europe sometime in the next decade or two.

In sum, a substantial increase in NATO conventional forces is unfeasible politically. Even if it could be achieved, it would not compensate for the decline in the credibility of nuclear retaliation as a deterrent. To be effective, deterrence has to move beyond the possibility of defense and include the probability of retaliation. Conventional deterrence requires not just an increase in conventional forces; it also requires a reconstitution of conventional strategy.

THE ROLE OF CONVENTIONAL RETALIATION

The new element required in NATO strategy is conventional retaliation. NATO has four possible means of deterring Soviet aggression: defense with conventional or nuclear forces and retaliation by conventional or nuclear forces. Under MC 14/3, NATO relied on a sequence of three responses: conventional defense, nuclear defense, nuclear retaliation (Figure 2-1). The decreasing credibility of NATO use of nuclear weapons, however, has loosened the connections between these responses (indicated by the dotted lines in Figure 2-2). As a result, both nuclear and retaliatory deterrence are weakened. The problem is to restore the latter without resorting to the former. The need, in short, is to add some form of conventional retaliation to NATO strategy (Figure 2-3). That retaliatory component can best take the form of provision for, in the case of a Soviet attack, a prompt conventional retaliatory offensive into Eastern Europe.

FIGURE 2-1. ORIGINAL FLEXIBLE RESPONSE		
MISSION	NUCLEAR FORCES	CONVENTIONAL FORCES
DEFENSE	3 ←————————————	1
RETALIATION	4	2
FIGURE 2-2. DETERIORATED FLEXIBLE RESPONSE		
MISSION	NUCLEAR FORCES	CONVENTIONAL FORCES
DEFENSE	3 ← — — — — — — — — —	1
RETALIATION	4	2
FIGURE 2-3. RECONSTITUTED FLEXIBLE RESPONSE		
MISSION	NUCLEAR FORCES	CONVENTIONAL FORCES
DEFENSE	3 ← — — — —	1
RETALIATION	4	— — 2

For the threat of retaliation to be an effective deterrent it must (a) be directed against a target that is highly valued by the potential aggressor and (b) have a high degree of probability it will be implemented. It is reasonable to assume that the Soviet elite values, next to the security of the Soviet Union itself, the security of its satellite regimes in Eastern Europe. If the threat of nuclear attacks against the Soviet Union has lost its credibility, the next most effective threat NATO can pose surely is the possibility of a conventional retaliatory offensive directed against the Soviet empire in Eastern Europe. In addition, as Snyder observed, the credibility of the threat of retaliation to "a large-scale Soviet ground attack on Western Europe depends on convincing the enemy that we would gain more by carrying out this threat than we would lose."[10] Precisely because this condition is no longer met, the threat of nuclear retaliation has lost its credibility. No such problem arises, however, by the threat of a conventional retaliatory offensive into Eastern Europe.

Almost every other form of retaliation against conventional attack involves escalation, either vertical, as in NATO doctrine, or, conceivably, horizontal. A conventional offensive into Eastern Europe, in contrast, is retaliation in kind, at the same level and in the same theater as the initial attack. It thus has unimpeachable credibility. Just as the Soviets have to believe that the United States would retaliate in kind against a strategic attack on American cities or a theater nuclear attack on Western Europe, they would also have to believe that the United States and its allies would retaliate in kind against a conventional attack on West Germany. Deterrence without retaliation is weak; retaliation through escalation is risky. Conventional retaliation strengthens the one without risking the other.

Strategy should exploit enemy weaknesses. A deterrent strategy that included provision for conventional retaliation would do this in two ways. First, it would capitalize on the uncertainties and fears that the Soviets have concerning the reliability of their Eastern allies, and the uncertainties and fears that the governments of those countries have concerning the reliability of their own peoples. It would put at potential risk the system of controls over Eastern Europe that the Soviets have developed over thirty years and which they consider critical to their own security. The deterrent impact of the threat of conventional retaliation would be further enhanced by prior Allied assurances to Eastern European governments that their countries would not be invaded if they abstained from the conflict and did not cooperate in the Soviet attack on the West. At the very least, such an invitation would create uneasiness, uncertainty, and divisiveness within satellite governments, and hence arouse concerns among the Soviets as to their reliability. In practice, the Allied offensive would have to be accompanied with carefully composed political-psychological warfare appeals to the peoples of Eastern Europe stressing that the Allies were not fighting them but the Soviets and urging them to cooperate with the advancing forces and to rally to the liberation of their countries from Soviet military occupation and political control. A conventional retaliatory strategy is based on the assumption that the West German reserves, territorial army, and populace will put up a more unified, comprehensive, and determined resistance to occupation by Soviet armies than the East German, Czech, Polish, and Hungarian forces and populations will to liberation from Soviet armies. (If this assumption is unwarranted, the foundations of not only a conventional retaliatory strategy but also of NATO would be in question.) Politically speaking, the Soviet Union has more to lose from Allied armies invading Eastern Europe than NATO has to lose from Soviet armies invading Western Europe. The Soviet Union should, consequently, give higher priority to preventing an Allied offensive into Eastern Europe than to pushing a Soviet offensive into Western Europe.

If the satellites did fight, the extent of their participation in a war, it is generally recognized, would depend on the scope and speed of Soviet success in the conflict. So long as the Soviets are moving westward, they are more likely to have complacent and cooperative allies. If, however, they are stalemated or turned back, disaffection is likely to appear within the Warsaw Pact. A prompt Allied offensive into Eastern Europe would stimulate that disaffection at the very start of the conflict. Neither the Soviets nor, more importantly, the satellite governments could view with equanimity West German tanks on the road to Leipzig and Berlin and American divisions heading for Prague and Cracow. From the viewpoint of deterrence, such a prospect would tremendously enhance the undesirability of war for the governments of these satellite countries. Those governments which provide more than one-third of the Warsaw Pact combat forces on the central front, would lose more than anyone else in such a war and hence would become a puissant lobby urging their Soviet partner not to initiate war.

A conventional offensive into Eastern Europe would thus threaten the Soviets where they are politically weak. It would also be aimed at Soviet military weakness. Both Western observers and Soviet military leaders agree that Soviet officers and NCOs are much better at implementing a carefully detailed plan of attack than they are at adjusting to rapidly changing circumstances. A conventional offensive into Eastern Europe would confront the Soviets with just exactly the situation their doctrine and strategy attempt to avoid; one in which they do not have control of developments and in which they face a high probability of uncertainty and surprise. It would put a premium on flexibility and adaptability, qualities in which the Soviets recognize themselves to be deficient. One knowledgeable observer has even argued that, ''If the Soviet Union were poised to launch an offensive, and were preempted in this by a NATO spoiling attack, there is little doubt that, in their own eyes, the Soviets reckon that *they* stand a good chance of collapse.''[11]

A prompt Allied offensive into Eastern Europe would also greatly increase the probability of a protracted war. Soviet planning, however, is in large part directed toward a short-war scenario in which the Soviets score a breakthrough, occupy a substantial portion of West Germany, and then negotiate a cease-fire from a position of strength. With a retaliatory strategy, Soviet armies might be in West Germany but Allied armies would also be in East Europe, and driving them out would require more time for mobilization and organization of a counteroffensive.

The basic point, moreover, is deterrence. The prospects for the sustained success of the Allied offensive into Eastern Europe do not have to be 100 percent. They simply have to be sufficiently better than zero and to raise sufficient unpleasant uncertainties to increase significantly the potential costs and risks to the Soviets of starting a war.

Current NATO strategy already contemplates the possibility of a counteroffensive. It would occur after the enemy's offensive forces have penetrated NATO territory and then been slowed or brought to a halt and NATO forces have been reinforced. A counteroffensive follows sequentially after the enemy's offensive and is directed to retrieving the initiative and recovering occupied territory.[12] A retaliatory offensive, in contrast, occurs simultaneously with the enemy's offensive. Its primary purpose is not to strike the enemy where he has further advanced, as is usually the case with a counteroffensive; rather it is to attack him in an entirely different sector. It thus would have a very different impact on Soviet force planning. The threat of a counteroffensive will lead the Soviets to make their offensive drive as strong as possible in order to advance as far as they can and do as much damage as they can to NATO's defensive forces and thus to postpone or blunt NATO's counteroffensive possibilities. The threat of a retaliatory offensive, on the other hand, will lead the Soviets to worry about their defensive capabilities and hence to deploy their forces more evenly across the entire front. A counteroffensive threat, in short, will lead the Soviets to strengthen their offense; a retaliatory offensive threat will lead them to weaken it.

NATO adoption of a conventional retaliatory option would thus pose a new problem to the Soviets. The Soviet military forces in Europe are now almost entirely offensively oriented. Soviet doctrine places overwhelming emphasis on the importance of the offensive and, in particular, on the need for both speed and surprise so as to achieve Soviet objectives before NATO reinforcements arrive and NATO decides to use nuclear weapons. At present the Soviets are free to develop their plans and forces for a lightning and overpowering offensive into Western Europe without having to worry about any defensive needs, other than air defense. The Soviets, as Richart Burt has observed, "have designed and trained a force to attack, not to defend. Whatever their ultimate plans, the Soviets have deployed their forces to seize territory, not to hold it."[13] They have been able to do this, however, only because NATO has permitted them to do so. NATO strategy has given the Soviet offensive a free ride. If, however, the Soviets also had to consider the possibility of a prompt NATO conventional offensive, they would either have to reallocate forces from offensive to defensive missions or to devote still more scarce resources to military purposes to meet this need.

The purpose of a conventional retaliatory option is to deter Soviet attack on Western Europe. The capability to exercise that option, however, could also contribute to the deterrence of Soviet aggressive moves in other parts of the world. At present the Soviets know that they could advance in force into the Persian Gulf area without having to worry about the security of their flank in central Europe. Their position is, in this respect, similar to that of Hitler in the 1930s. Although France had various commitments to Poland and the Little Entente, which presupposed, as DeGaulle argued, an offensively oriented army, it could not in fact pose any deterrent threat against Hitler's moving eastward because it had adopted a purely defensive strategy, symbolized by the Maginot Line. French military strategy left Hitler free to do what he wanted in Eastern Europe. In similar fashion, NATO forces do not now pose even a theoretical restraint on Soviet moves elsewhere. If, however, NATO were prepared to launch a military offensive into Eastern Europe, the Soviets would have to assure themselves as to the adequacy of their defenses there and as to the loyalty of their allies before they could take the offensive against Iran, Pakistan, China, Japan, or any other neighboring state.[14]

The point is sometimes made that NATO is a defensive alliance and that a defensive alliance requires a defensive strategy. This argument has no basis in logic or history. NATO is a defensive alliance politically, which means that its purpose is to protect its members against Soviet attack through deterrence if possible and through defense if necessary. There is, however, no reason why a politically defensive alliance cannot have a militarily offensive strategy. Such a strategy may, indeed, be essential to securing the deterrent purposes of the alliance. For two decades NATO did in fact pursue its purposes primarily through the threat of launching a strategic nuclear offensive against the Soviet Union. If a nuclear offensive is compatible with the defensive purposes of the Alliance, certainly a conventional offensive should be also. Given long-standing NATO reliance on the possible first use of tactical nuclear weapons and, if necessary, strategic retaliation against the Soviet Union itself, it would be rather anomalous for its members to find something unduly abhorrent about a conventional offensive into Eastern Europe. On moral and political grounds, surely it is far more desirable to deter by threatening to liberate Eastern Europeans than by threatening to incinerate Russians.

THE MILITARY FEASIBILITY OF CONVENTIONAL RETALIATION

At this point, the reader may well be saying to himself: "Your argument is all wonderful in theory, *but* (a) as you've pointed out,

NATO is not meeting its own already-established conventional buildup goals, and (b) the strategy you advocate would require a buildup far larger than anything NATO has contemplated. Conventional retaliation just is not practical." The question is: What are the military requirements of a conventional retaliatory offensive?

The answer is not as great as one might think.

First, it is necessary to clear away the popular cliche that the offensive requires a three-to-one overall superiority. If this were the case, NATO's problems would be over. Under no circumstances, given the current balance and probable rates of mobilization on each side, could the Warsaw Pact achieve an overall three-to-one superiority over NATO. Most scenarios do not deviate much from Robert Fischer's 1976 estimate that Pact superiority in men in combat units would peak at about 2:1 two weeks after Pact mobilization began, assuming NATO mobilization lagged one week.[15] Unfortunately, however, 3:1 overall superiority is not what is required to attack. It is instead what may be required at the exact point of attack. Achieving that superiority is the product not of overall superiority in numbers but rather of superiority in mobility, concentration of forces, deception, and surprise.

Second, while the Soviets clearly do have a significant conventional superiority in Europe in numerical terms, that superiority is not enough, in itself, to give them a decisive advantage. In 1981, Pact superiority in divisional manpower was roughly 1.36:1, but in terms of overall manpower there was almost equality, with a ratio of 1.09:1. The Pact had many more tanks than NATO, but NATO was better off in attack aircraft. In terms of armored division equivalents (ADEs), perhaps the single most useful measure, the ratio was 1.2:1. Overall the Pact wins the numerical bean count, but it does not have an advantage which would guarantee victory in war.[16] If a high probability currently exists of the Soviets' achieving substantial success in a central European war, that stems as much from their strategy as from their numbers. They are planning to concentrate their forces and use them offensively in the most militarily effective manner possible, while NATO has, for a variety of reasons, been committed to a defensive strategy which almost ensures military defeat.

Third, a force which is inferior in overall strength can still pursue an offensive strategy. History is full of successful examples. The

German offensive into France in 1940 and the North Vietnamese offensive in 1975 are two such cases. As US Army FM 100-5 points out, other examples are the Third Army's attack through France in 1944, the US offensive in Korea in 1951, and the Israeli Sinai campaign of 1967. In these cases, as in Grant's Vicksburg campaign (cited at length in FM 100-5 as a model offensive), the attackers succeeded "by massing unexpectedly where they could achieve a brief local superiority and by preserving their initial advantage through relentless exploitation."[17]

Obviously, stronger forces are more desirable than weaker ones. Implementing a strategy that includes conventional retaliation, however, requires more changes in the NATO military mind-set than it does in NATO military forces. For thirty years NATO has thought about conventional warfare exclusively in defensive terms. It has assumed that all the ground war and the bulk of the war generally would be fought in West Germany. It has pursued a strategy of forward defense, entirely defensible and necessary in terms of German interests, that leaves NATO forces strung out along the entire eastern border of the Federal Republic and hence highly vulnerable to an overpowering Soviet concentration of offensive forces. It has, moreover, done this without being able, also for understandable political reasons, to construct major fortifications that could slow down and greatly complicate a Soviet attack. NATO developed, as *The Economist* put it, "a Maginot-line mentality without the Maginot line."[18]

Fortunately there are signs that this mentality may be changing. Supreme Headquarters Allied Powers Europe (SHAPE) is developing plans for the deep interdiction of Warsaw Pact second-echelon forces. The aim is to locate Pact follow-on forces through improved intelligence and to attack them with long-range conventional means before they reach the battle zone, while at the same time NATO forces are holding the forward defense line against Pact first-echelon forces. It is, as General Rogers said, a way of adding "depth to the battlefield by extending the area of NATO's operations into the *enemy's* rear area."[19] A conventional retaliatory offensive as proposed here is compatible with and would supplement this emphasis on deep interdiction. It would involve NATO operations into the enemy's rear at the operational rather than simply the tactical level. It would employ not just conventional PGMs and missiles but the full range of conventional combined arms, and it would also serve to disrupt enemy logistics and reinforcements. Similarly, a retaliatory offensive is highly compatible with US Army AirLand Battle doctrine, with its emphasis on the initiative, deep attack, and maneuver: *"Initiative, the ability to set the terms of the battle by action, is the greatest advantage in war . . . . The offense is the decisive form of war, the commander's only means of attaining a positive goal or of completely destroying the enemy force."*[20] There are at least some signs that German military thinking may be moving in a similar direction.[21]

A strategy with an offensive component would better capitalize on the current capabilities of NATO forces in Europe than does a purely defensive strategy. By and large, these forces are heavy forces; two-thirds of the Allied divisions in Germany are armor divisions; most of the rest are mechanized infantry. It is often said, of course, that these forces will enable NATO to have a mobile defense and to launch counteroffensives. That is true, and the same qualities also make them suited for a retaliatory offense. It is a misuse of expensive resources to consign these heavy forces primarily to a defensive role. In addition, NATO's forward defense strategy has always caused problems with respect to how Allied forces in the various sectors could reinforce each other. If the Soviets, for instance, launch their principal attack across the North German plain, what role could the substantial American and German forces in southern Germany play in bringing that advance to a halt? To move those forces laterally, that is parallel to the front, would be a logistical nightmare and could leave Bavaria open to a secondary Soviet attack. Not to move those forces northward on the other hand, would greatly facilitate the Soviets' overwhelming the NATO forces in the north. The most efficient use of any substantial Allied forces not close to the Soviet attack corridors is to carry the war to the enemy.

The solution to NATO's deterrence problem is not to be found in any particular technological or doctrinal gimmick. It requires a diversified effort including more resources, qualitative improvements, and strategic innovations. Preparing for a retaliatory offensive will not do it alone, but it cannot be done without preparing for a retaliatory offensive.

In practical terms, what might a retaliatory offensive look like?

If the threat of such an offensive is to serve its deterrent purpose, the Soviets must have good reason to believe that an offensive is possible and little knowledge as to exactly where and when it might occur. NATO military planning for such an offensive would have to encompass a variety of alternative scenarios and possible options reflecting Warsaw Pact deployments and axes of advance; NATO force capabilities and deployments; and East European politics, which might dictate withholding or limiting NATO offensive actions. It is clearly not possible to spell out in this paper detailed plans for a NATO offensive. Many different possibilities exist. To give some idea as to what could be involved, however, it might be desirable briefly to elaborate what is undoubtedly the most obvious scenario for both Soviet and NATO planners. Because it is the most obvious scenario, it could also be one which is unlikely to be realized in practice.

Three of the most probable Soviet invasion routes are across the North German plain to Hannover and then northward towards Bremen and Hamburg, through the Gottingen corridor towards the Ruhr, and through the Fulda Gap towards Frankfurt. These attacks would be led by the powerful Third Shock Army (which includes four tank divisions and one motorizied rifle [MR] division) in the center of the front and the 2nd Guards Tank Army (one tank and two MR divisions) to the north.[22] They would engage the Dutch, German, British, and Belgian forces in NORTHAG and the III German and V American Corps in CENTAG. In these circumstances, the most appropriate retaliatory offensive would be by the VII American and II German Corps plus the 12th Panzer Division from the III German Corps. These are among the strongest and best equipped Allied forces in Central Europe; their Leopard II and Abrams tanks would provide the heart of the offensive thrust.

The offensive could well consist of two prongs. The major thrust would be through the Hof corridor towards Jena and Leipzig (see Figure 2-4). Its primary axis of advance would not be west-east but rather south-north, and hence the problem of river barriers would be minimized. The Soviet forces immediately on the scene include the three motorized rifle divisions and one tank division of the 8th Guards Army, headquartered in Weimar. Such an offensive would threaten the most direct Soviet supply routes supporting their forces in the Fulda Gap. The second prong would be launched in a more easterly direction towards Karlovy Vary and Teplice in Czechoslovakia. The immediate Soviet resistance would come from

FIGURE 2-4. POSSIBLE SOVIET OFFENSIVES AND NATO RETALIATORY OFFENSIVES

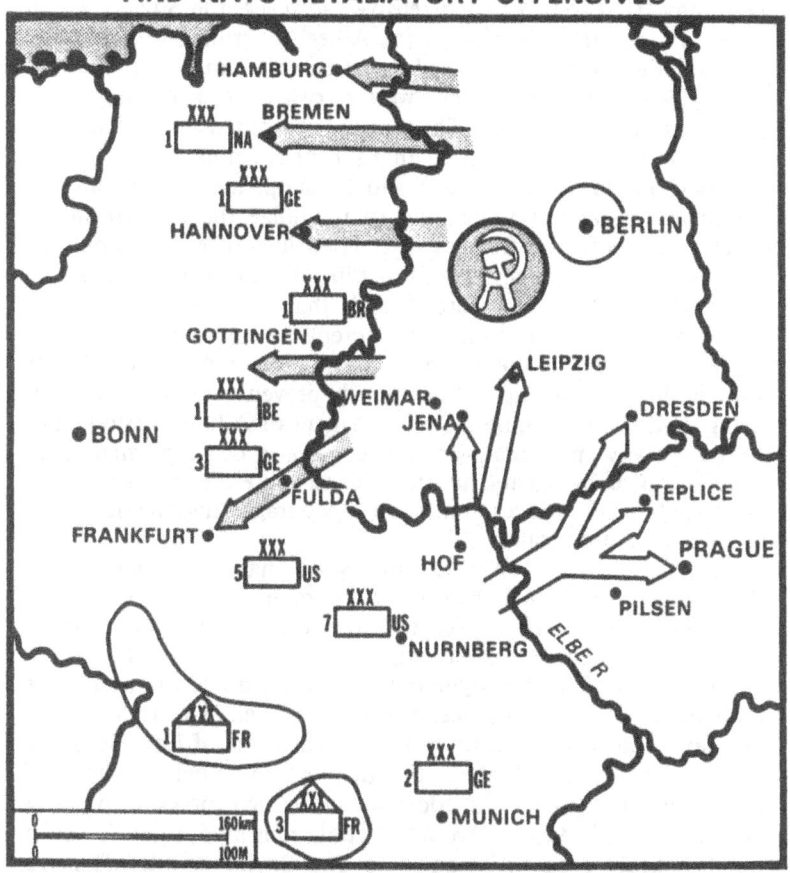

a single division deployed north of Pilsen. If this advance reached the Elbe, it could then either swing north towards Dresden or south towards Prague. The second prong would also help protect the southern and eastern flanks of the main prong.

The Allied forces engaged in these offensive moves would be superior in manpower, tanks, and ADEs to the Soviet forces immediately deployed against them. To the north and east of the 8th Guards Army, however, is the 1st Guards Tank Army. It is roughly comparable to the 3rd Shock Army and clearly is designed to make a major contribution to the Soviet offensive into West Germany. If it joined that offensive, however, the Soviets would have to face the possibility of the Allies' overrunning their other forces in the south. If they used the 1st Guards Tank Army to blunt the Allied offensive, they would risk not achieving their breakthrough in the north. The purpose of a retaliatory offensive is to confront them with precisely that sort of dilemma.

Allied military dispositions should supplement political and diplomatic measures in helping to minimize the enthusiasm of satellite forces for the Soviet cause. The Allied offensive should be directed at Soviet forces. The thrust into East Germany should be primarily by German forces and that into Czechoslovakia exclusively by American ones. At present the deployment of Allied forces in Bavaria is not the most satisfactory from this point of view, although it would not necessarily prevent Allied forces from being used in this manner. The movement of Allied divisions into East Germany and Czechoslovakia could also be supplemented by the infiltration by sea and air into Poland and Hungary of specially trained Special Forces units to encourage disaffection and resistance in those countries.

For many years NATO strategists have bemoaned the deployment of Allied forces in Germany, a legacy of the occupation, which left US forces in the south, some distance from the highly probable Soviet axis of advance across the North German Plain. This deployment is, however, made to order for a retaliatory offensive. It places US forces as well as German forces in a favorable location for a move into the heart of East Germany which would be highly threatening to Soviet lines of communication (LOCs). In addition, the French forces in southern Germany constitute a reserve which could reinforce Allied forces in the Frankfurt area or respond to any Soviet counteroffensive against Germany, e.g., through Austria.

How successful would be a retaliatory offensive such as this? That clearly would depend, among other things on:

• the size, character, and leadership of the NATO forces committed to the offensive;

• the strength and readiness of the opposing Warsaw Pact forces;

- the degree of surprise NATO achieved; and,
- the extent to which non-Soviet Warsaw Pact forces fought vigorously alongside their Soviet allies.

Just how these factors would play out is impossible to predict in advance. At one extreme, it is conceivable although unlikely that NATO forces could sweep north towards the Baltic and join up with amphibious forces in a giant pincer movement cutting East Germany in half and isolating Soviet forces to the west. At the other extreme, they might penetrate only a few kilometers into East Germany and Czechoslovakia. The point is that neither side could know for sure in advance, and that uncertainty is precisely what is required to reinforce deterrence. The Soviets would only know that, if they went to war under these circumstances, they would be putting at risk far more of great value than they would be at present.

Some changes in NATO forces are desirable to enhance the feasibility of a conventional offensive. These would not necessarily involve expansion in force levels. They would, however, require:

- redeployment of German forces in II Corps to positions closer to the inter-German border and compensating movement of some US forces to positions on the Czech border;
- major improvements, already called for in NATO plans, in stockpiles of fuel, ammunition, spare parts, equipment, and other supplies necessary to sustain an offensive movement; and,
- emphasis in weapons procurement on those items most relevant to offensive needs, e.g., attack aircraft, attack helicopters, long-range PGMs.

In addition to these actions, it would also be highly desirable to strengthen NATO defensive capabilities through the construction of fortifications and improvement in West German reserves and territorial forces. At present NATO follows a forward defense strategy but lacks forward fortifications. The principal reason for this has been the reluctance of the West German government to create a major fortified line that would give concrete embodiment to a permanent division of Germany. If, however, NATO strategy included provision for the invasion and liberation of at least portions of East Germany, a fortified line along the inter-German border would no longer have the symbolism that the Bonn government fears. The construction of such a line would, of course, make it possible to release additional Allied forces to offensive missions. The same result will also be achieved to the extent that territorial army units play a larger role in area defense.[23]

The special requirements for a conventional offensive capability will obviously compete with other claims on the modest and only slowly growing NATO resource. The central criterion for allocating resources among these competing claims, however, should be the extent to which they contribute to deterrence. Improvements in NATO defensive capabilities strengthen deterrence, but only marginally so. At the most they simply require the Soviets to invest comparable additional resources in their forces so as to maintain the same probability of success. Enhancement of NATO offensive capabilities, on the other hand, confronts the Soviets with an entirely new danger in terms of Allied penetration into and disruption of their Eastern European empire. It thus forces them to set this risk off against the advantages of attacking NATO and to mobilize or to divert resources from other sources if they are to meet that threat. A given increase in NATO offensive capabilities, in short, will produce a considerably higher return in terms of deterrence than the same investment in defensive capabilities. As a result, this addition to NATO strategy will also lower the total new resources NATO needs to invest to achieve effective deterrence. It could make conventional deterrence not only more credible but also cheaper than it would otherwise be.

THE POLITICS OF CONVENTIONAL RETALIATION

If conventional retaliation is strategically desirable and militarily feasible, the final question is whether it is politically possible. Will the Alliance agree to this amendment to the long-standing strategic doctrines set forth in MC 14/3?

Some may say that this proposal involves a fundamental change in NATO strategy for which it will be difficult if not impossible to mobilize support within the Alliance. In fact, however, incorporation of a conventional retaliatory offensive into NATO's strategy would, in many respects, be less a change in strategy than an effort to prevent a change in strategy. As it is, flexible response is inexorably becoming a dead letter. NATO strategy is changing fundamentally from a multi-pronged flexible response to a single-prong conventional defense. The addition of a conventional retaliatory option would, as the figures on page 23 indicate, simply restore some element of flexibility to a strategy that is rapidly becoming flexible. It would pose new uncertainties for the Soviet Union. It would adapt flexible response to the conditions of the 1980s. In similar fashion, a retaliatory offensive is not incompatible with the idea of forward defense. The latter is a necessary and appropriate response to German concern that as little of their country as possible become the locus of battle and subject to Soviet occupation. A retaliatory offensive would move at least some of the battle from West Germany to East Germany and Czechoslovakia. It is thus not a substitute for a strategy of flexible response and forward defense, broadly conceived, but rather a fleshing out of that strategy in changed circumstances. It would, in effect, make flexible response more flexible and forward defense more forward.

Adoption of a conventional retaliatory option would reinvigorate flexible response through conscious choice. Inevitably, the political feasibility of making such a conscious choice is a function of time and circumstance. In democratic societies, no new policy suggestion is immediately feasible. Every new proposal has to go through a process of discussion, consideration, analysis, amendment, and often initial rejection before it becomes reality. This is true in military policy as well as domestic policy, and it is doubly true in *alliance* military policy. The changes in NATO strategy in the mid-1950s and in the mid-1960s each required about five years to be implemented. There is no reason to think that the time required for change in the mid-1980s will be much different, nor to think that such change will not occur.

A broad consensus already exists on the need to enhance conventional deterrence. The political support of NATO governments and peoples for moving in this direction will in due course emerge, as is true of any new policy, from consideration of the unpalatability of the alternatives. In this instance, there are, broadly speaking, two such alternatives. One is to acquiesce in a greatly weakened deterrent, as the credibility of a US nuclear response declines. This is, in terms of short-range politics, unquestionably the easiest way out, but it is one which also will have its political costs in terms of both heightened Soviet influence over alliance members and heightened political tensions among alliance members. The other alternative to effective conventional deterrence is to recreate a credible nuclear deterrent. Nuclear deterrence of a conventional attack is most credible—and, indeed, may *only* be credible—when the national existence of the deterring state is at risk. No one doubts that an Israeli government would use nuclear weapons to prevent Arab armies from overrunning Tel Aviv. A similar rationale furnishes the explicit justification for the French nuclear force and the implicit justification for the British one. Nuclear deterrence could be restored in central Europe if an independent, invulnerable, modest-sized German nuclear force were brought into being. The Soviets would have to believe that such a force would be used in the event of a Soviet attack on the Federal Republic. The political problems involved in the creation of such a force, however, dwarf those that arise from adoption of a conventional retaliation strategy.[24]

The strategic environment in the United States is increasingly favorable towards conventional retaliation becoming a NATO option. The other key locus of decisionmaking is the Federal Republic. One would think that German leaders would endorse a military strategy that, in comparison to the alternatives, promised to produce stronger deterrence at lower cost, to reduce the probability that nuclear weapons would be used in the territory of the Federal Republic, and to shift at least some of the fighting, if war did occur, from the Federal Republic to East Germany and Czechoslovakia. It is hard to see why it might be good politics in West Germany to oppose such a move. If, after the normal debate necessary for policy innovation in any democratic country, the West German government was unwilling to support such a change, the United States would clearly have to reconsider its commitment of forces to a strategy and posture that is doomed to be found wanting. "For deterrence to be credible," as General Rogers has said, "it requires capabilities adequate for successful defense and effective retaliation."[25] Effective retaliation means credible retaliation, and, in today's world, credible retaliation means conventional retaliation. That is the inescapable logic that will drive NATO's strategic choices in this decade.

ENDNOTES

1. Henry A. Kissinger, "The Future of NATO," in *NATO: The Next Thirty Years*, ed. Kenneth A. Meyers (Boulder, Colo.: Westview Press, 1980), p. 7

2. McGeorge Bundy, "Strategic Deterrence Thirty Years Later—What Has Changed?" in *The Future of Strategic Deterrence, Part I*, Adelphi Papers, No. 160 (London: International Institute for Strategic Studies, 1980), pp. 10-11: ". . . the strategic protection of Europe is as strong or as weak as the American strategic guarantee. . . . the effectiveness of this American guarantee is likely to be just as great in the future as in the past." Cf. McGeorge Bundy, George F. Kennan, Robert S. McNamara, and Gerard Smith, "Nuclear Weapons and the Atlantic Alliance," *Foreign Affairs*, vol. 60, no. 4 (Spring 1982), pp. 754, 765: U.S. "willingness to be the first . . . to use nuclear weapons to defend against aggression in Europe . . . needs reexamination now. Both its cost to the coherence of the Alliance and its threat to the safety of the world are rising while its deterrent credibility declines. . . . [T]he present unbalanced reliance on nuclear weapons, if long continued, might produce [some deeply destabilizing]political change."

3. Leo P. Crespi, "West European Perceptions of the U.S." (Paper presented at the International Society of Political Psychology Convention, June 1982), Table 4, quoted in Bruce Russett and Donald R. Deluca, "Theater Nuclear Forces: Public Opinion in Western Europe," *Political Science Quarterly*, vol. 98, no. 2 (Summer 1983), p. 194.

4. CBS/*New York Times* poll, May 1982, cited in "Opinion Roundup," *Public Opinion*, vol. 5 (August-September 1982), p. 38.

5. General Bernard W. Rogers, "Greater Flexibility for NATO's Flexible Response," *Strategic Review*, vol. 11, no. 2 (Spring 1983), pp. 11-19; Michael Howard, "Reassurance and Deterrence: Western Defense in the 1980s," *Foreign Affairs*, vol. 61, no. 3 (Winter 1982-83), pp. 309-343; Senator Sam Nunn, "NATO: Can the Alliance Be Saved?" Report to the Senate Committee on the Armed Services, 97th Cong., 2d sess., May 13, 1982; "Former Defense Chiefs Urge Allies to Alter Conventional, Nuclear Policy," *The Washington Post*, February 2, 1983, p. 10; *Strengthening Deterrence in Europe: Proposals for the 1980s*, Report of the European Security Study (New York: St. Martin's Press, 1983); Bundy et. al., "Nuclear Weapons and the Atlantic Alliance," pp. 753-768; Eckhard Lubkemeier, "Problems, Prerequisites, and Prospects of Conventionalizing NATO's Strategy," (Unpublished paper, Bonn, Friedrich-Ebert-Stiftung, February 1983); English version, July 1983; pp. 3 and *passim*.

6. Glenn H. Snyder, *Deterrence and Defense: Toward a Theory of National Security* (Princeton, N.J.: Princeton University Press, 1961), pp. 4, 14-16.

7. John J. Mearsheimer, *Conventional Deterrence* (Ithaca, N.Y.: Cornell University Press, 1983), pp. 19-20. Mearsheimer identifies six additional possible cases of conventional deterrence, in five of which deterrence failed.

8. Saadia Amiel, "Deterrence by Conventional Forces," *Survival*, vol. 20, no. 2 (March-April 1978), p. 59.

9. For analyses in depth of the failure of pre-World War II deterrence, see Mearsheimer, *Conventional Deterrence*, esp. chapters 3, 4, and Scott D. Sagan, "Deterrence and Decision: An Historical Critique of Modern Deterrence Theory" (Ph.D. diss., Harvard University, 1983), chapters 4-7. The bomber introduced the possibility of more effective deterrence by retaliation and compelled aggressors to

take countermeasures. Both Hitler and the Japanese were deeply worried about retaliatory air attacks on their cities. Hitler guarded against this by mutual deterrence through an informal understanding with the Allies that neither side would target population; Japan guarded against it by a disarming first strike. Neither Axis power was deterred by Allied defensive measures, including the Maginot Line.

10. Snyder, *Defense and Deterrence,* p. 79.

11. Christopher N. Donnelly, "Soviet Operational Concepts in the 1980s," in European Security Study, *Strengthening Conventional Deterrence,* p. 135. See also the report of the Steering Group of this study, p. 18, and Joshua M. Epstein, "Soviet Confidence and Conventional Deterrence in Europe" (Unpublished paper, Harvard University, Center for International Affairs, 1982.)

12. See, for example, Richard B. Remnek, "A Possible Fallback Counteroffensive Option in a European War," *Air University Review,* vol. 35, no. 1 (November-December 1983), pp. 52-62.

13. Richard Burt, "The Alliance at a Crossroad" (Address, Friedrich-Ebert-Stiftung, Bonn, December 2, 1981). See also European Security Study, *Strengthening Conventional Deterrence,* p. 55.

14. For elaboration of this point, see Huntington, "The Renewal of Strategy," in *The Strategic Imperative: New Policies for American Security,* ed. Huntington (Cambridge, Mass.: Ballinger, 1982), pp. 24-29.

15. Robert Lucas Fischer, *Defending the Central Front: The Balance of Forces,* Adelphi Papers, No. 127 (London: International Institute for Strategic Studies, 1976), pp. 24-25.

16. For two excellent analyses emphasizing the uncertainties on the Central Front, see John J. Mearsheimer, "Why the Soviets Can't Win Quickly in Central Europe," *International Security,* vol. 7, no. 1 (Summer 1982), pp. 5-39, and Anthony H. Cordesman, "The NATO Central Front and the Balance of Uncertainty," *Armed Forces Journal International,* vol. 120, no. 12 (July 1983), pp. 18-58.

17. US Army, FM 100-5, *Operations* (August 20, 1982), p. 8-5.

18. "Thrust Counter-Thrust," *The Economist,* May 1, 1981, p. 15: "NATO . . . should plan an armoured counter-thrust towards the Warsaw pact's rear areas from the part of the front that is not under communist attack. The knowledge that such a counter-offensive was part of NATO's strategy would, at the very least, complicate the Soviet plan of attack. It could even prevent the attack from happening. The Warsaw pact should not have the luxury of thinking that attack is its monopoly."

19. Rogers, "Greater Flexibility for NATO's Flexible Response," p. 17.

20. FM 100-5, pp. 7-2, 8-1.

21. See K.-Peter Stratmann, "Prospective Tasks and Capabilities Required for NATO's Conventional Forces," in European Security Study, *Strengthening Conventional Deterrence,* p. 163.

22. Information on the deployments and strengths of Soviet and Allied forces comes from: Fischer, *Defending the Central Front;* Cordesman, "The NATO Central Front," pp. 18ff.; and John Erickson, *Soviet-Warsaw Pact Force Levels* (Washington: United States Strategic Institute, Report 76-2, 1976). For simplicity's sake, the discussion here is couched purely in terms of deployed forces, without reference to the reinforcements possible on both sides. In addition, the data used are meant to be illustrative, not necessarily definitively accurate.

23. See, for example, the supportive but contrasting views of Steven L. Canby, "Territorial Defense in Central Europe," *Armed Forces and Society,* vol. 7, no. 1 (Fall 1980), pp. 51-67, and Waldo L. Freeman, Jr., *NATO Central Region Forward Defense: Correcting the Strategy/Force Mismatch* (Washington: National Defense University Research Directorate, National Security Affairs Issue Paper Series 81-3, 1981).

24. For further discussion of this option, see Huntington, "Broadening the Strategic Focus," in *Defense and Consensus: The Domestic Aspects of Western Security, Part III,* Adelphi Papers, No. 184 (London: International Institute for Strategic Studies, 1983).

25. Rogers, "Greater Flexibility for NATO's Flexible Response," p. 15.

CHAPTER 3

STRATEGIC IMPLICATIONS OF DOCTRINAL CHANGE: A CASE ANALYSIS

by

Richard Hart Sinnreich

In *The Guns of August,* in a chapter entitled "The Shadow of Sedan," historian Barbara Tuchman describes the roots of 1914 French military strategy in the following words:

> Living in the shadow of that unfinished business, [her defeat in 1870] France, reviving in spirit and strength, grew weary of being eternally on guard, eternally exhorted by her leaders to defend herself. As the century turned, her spirit rebelled against thirty years of the defensive, with its implied avowal of inferiority. . . . She needed some weapon that Germany lacked to give herself confidence in her survival. The 'idea with a sword' fulfilled the need. . . .
>
> Translated into military terms . . . [it] became the doctrine of the offensive.[1]

Tuchman goes on to describe a doctrinal evolution in which tactical and strategic realities gradually gave way to a sort of military metaphysics, in which words like "initiative," "elan," and "the will to win" came to assume a transcendental character entirely disconnected from France's military circumstances, and in which the spirit of the French soldier became the talisman whose magic would destroy the "myth" of German invincibility. "Like the youth who set out for the mountaintop under the banner marked 'Excelsior!' the French Army marched to war in 1914 under a banner marked 'Cran' [guts]."[2] The trip cost France a generation of young men, and very nearly cost her the war.

History rarely repeats itself. But the processes which lead armies and nations to disaster do repeat themselves, with depressing regularity. There are indications—scattered but worrisome—that the same processes of military self-deception which nearly destroyed France in 1914 are today at work in the United States. They arise in part out of a doctrinal evolution which in many respects parallels the French experience—an evolution as necessary as it was inevitable, but which now as before threatens to go beyond the bounds of strategic prudence.

AN ARMY IN TRANSITION

Doctrine is only partly a reflection of military requirements. It is also very much a reflection of the army which develops it. To understand the evolution of US Army doctrine during the last twelve years, one must look first at the army in which it took place.

To begin with, like the French Army of 1870, the American Army of 1972 was a defeated army. Some will object that the Army in Vietnam never lost a major engagement. But the bottom line remains that the war was lost—a war against a nominally inferior opponent, and which for ten years consumed the lives, energies, and professional reputations of a generation of American soldiers. The trauma was commensurately severe.

Like its French predecessor, the US Army of 1972 was also an internally fractured army. Munity, desertion, and Dreyfus all had their latter-day counterparts in draft evasion, fragging, drugs, falsified body counts, and Calley. Article after article, book after book documented an army in serious trouble. It is almost impossible to believe, today, that in 1972, a book entitled *The Death of the Army* could be taken seriously.[3]

43

Psychological demoralization was accompanied by intellectual impoverishment. The Vietnam War cost the Army a decade of tactical literacy. As the war pulled year-group after year-group of officers into repetitive Vietnam tours, the professional development process was badly distorted. From West Point and ROTC through branch courses, Command and General Staff College, and the War Colleges, education in the art of war gave way in large measure to preparation for service in a single war—the only war we had. Even the Army's long-standing preoccupation with Europe could not survive unscathed a decade which saw US Army, Europe cannibalized to support Vietnam, and residual deployed forces stressed just to maintain internal discipline and cohesion. Once again it is instructive to recall the published literature: in 1973, two highly-regarded serving officers published a book which seriously proposed that the Army in Europe be postured primarily to facilitate its evacuation in the event of war.[4] No clearer or sadder evidence of the Army's loss of tactical self-confidence could be conceived, than that Dunkirk could thus be elevated to the status of a force design concept.

And yet it was not all that surprising, for like the nation itself, the Army of 1972 was an army which had lost its strategic bearings. With counterinsurgency bankrupt and the nation in strategic retreat, the Army found itself temporarily without an enemy—at least, without an external enemy. There were plenty of internal foes. Thus, in a directed study of future Army roles which was completed in 1972, the Army War College highlighted domestic civil action; while in a meeting late in 1971 with the new Secretary of the Army, instructors at West Point beseeched the Secretary for a definitive statement of the Army's post-Vietnam mission.[5]

DOCTRINAL DEBATE AND THE ACTIVE DEFENSE

It is in this last respect, of course, that the French case differed most markedly from our own. For France after 1870, neither the enemy nor the mission changed. But the practical consequences were the same. For in the effort to recover its sense of identity and purpose, the US Army turned from Vietnam back to its traditional focus on Europe. And what it saw there was not reassuring. During the intervening ten years, and assisted by the Czechoslovakian invasion of 1968, the conventional power of the Warsaw Pact had

44

increased radically. To an already crushing numerical preponderance, the Pact added qualitative improvements in virtually every category of equipment—armor, artillery, engineer, transport, and tactical aviation—in the process converting a ponderous, relatively inflexible conventional force structure into a formidable offensive capability.

Nor was the improvement limited to conventional forces. By 1972, the USSR was beginning to field tactical nuclear capabilities fully as diverse as our own, rapidly eroding the one clear advantage possessed by NATO, together with chemical offensive and defensive capabilities for which no Western counterparts even existed.

In the face of this staggering augmentation of offensive power, the US Army in the early 1970s found itself radically reduced in numbers, without a peacetime draft, and on the short end of a ten-year deficit in equipment modernization. The Army accordingly did precisely what its French predecessor had done in similar circumstances: it figuratively jumped into its foxholes. In doctrinal terms, it went on the defensive. It was encouraged no little in this reaction by the 1973 Arab-Israeli War, whose violence and lethality shocked officers conditioned by ten years of low-intensity warfare; and by the growing prominence of analytical and gaming techniques whose emphasis on the quantifiable components of combat power merely magnified the apparent force disparities with which the Army would have to cope in a war in Central Europe.

The formal expression of this response was a tactical concept known popularly—and not altogether accurately—as the "Active Defense." Promulgated in the 1976 edition of FM 100-5, *Operations,* the new doctrine in effect translated NATO's forward defense requirement into operational method, thus, like French doctrine after 1870, making a tactical virtue of what appeared to be strategic necessity.

But whereas in France, the subsequent revulsion against a defensive doctrine required thirty years to mature, in the US Army it developed almost immediately. Stimulated by critics skeptical of the premises of the Active Defense and concerned about its implications for Army modernization, debate over FM 100-5 began even before the ink dried on the coordinating draft. The debate continued for nearly five years, and as with the similar debate over French Army doctrine from 1911 to 1913, eventually expanded outside the cloistered walls of Fort Leavenworth and Fort Monroe to include civilian defense analysts, legislators, and even public commentators. The very scope of the controversy alternately bemused and disturbed Army leaders unaccustomed to seeing tactical arcana debated on the Op-Ed pages of *The New York Times*.[6]

The criticisms were as diverse as the critics. The Active Defense was assailed for its disregard of tactical imponderables like initiative, morale, and the offensive spirit; for its apparent exaltation of firepower over maneuver; and for its preoccupation with calculations of relative attrition to the exclusion of tactical creativity and judgment. All these criticisms had merit, and in all, there were resonances of the French doctrinal debate of 1911-13.

But the most telling criticism of the Active Defense was also the most significant indicator of the psychic condition of the Army by 1976: that the Active Defense was devoid of operational content. At best, critics charged, it promised only to defer defeat. It offered no avenue to operational success. In seeking to fulfill its doctrinal commitment to "winning the first battle," the Army was accused of becoming so preoccupied with fighting the first battle that it forgot all about winning the last. For an Army traumatized by ten years of tactical success culminating in operational failure, no critique could have been more devastating.

DOCTRINAL REVISIONISTS

As reaction against the 1976 doctrine intensified, several related developments occurred to accelerate the movement toward doctrinal revision.

The first was a more sophisticated understanding of the Soviet conventional threat. As analysts probed the reasons for the invariable collapse of the forward conventional defense in games and exercises, it became apparent that a major cause was the Warsaw Pact's ability to hold significant forces out of the forward battle for subsequent commitment against weaknesses developed or uncovered by the initial assault. In part, of course, this capability simply reflected the Warsaw Pact's numerical advantage. But its effects were magnified by the prescriptions of the 1976 doctrine, which discouraged the retention of reserves, and which instead proposed to react to enemy penetrations by lateral reinforcement from less threatened sectors. The result of this process, critics pointed out, even assuming it could be executed, was to present precisely the sort of opportunities which Soviet follow-on forces were designed to exploit.

Clearly, something had to be done to diminish the enemy's freedom to commit follow-on forces into the battle at times and places of his choosing. The incentive to do so was further increased by a growing recognition that the Soviet operational pattern presented weaknesses of its own—unprotected flanks, vulnerable lines of communication, and a command-and-control system whose centralization gave it great resilience only so long as events were unfolding according to plan. All of these vulnerabilities presented opportunities for dislocation and destruction through offensive countermaneuver, but only if something could be done to delay or disrupt the closure of enemy follow-on forces.

In the meantime, another line of inquiry was tending in the same direction, but from a quite different starting point. For as the relative US advantage in theater nuclear capabilities declined, so did confidence in both the credibility of existing employment doctrine and the Army's ability to operate on a nuclear or nuclear/chemical battlefield.

Accordingly, at the same time pressures were mounting to rethink the Active Defense, similar pressure arose to "get serious" about the nuclear-chemical business. Pursued under the rubric of "The Integrated Battlefield," getting serious comprised efforts ranging from the rehabilitation of the Army's dilapidated chemical warfare capability to the hardening of command, control, and communications systems against electromagnetic pulse degradation. For the purposes of this discussion, however, the most important aspect of "The Integrated Battlefield" was a fresh look at tactical nuclear employment doctrine.

Tactical nuclear employment is a complex issue. Suffice it here to say that as analysts both in the Army and in NATO began to reexamine tactical nuclear employment, an unexpected convergence began to emerge between the conventional and nuclear problems. Operationally, a nuclear employment doctrine must help maintain the forward defense; strategically, it must send a clear but controlled escalatory signal; and politically, it must neither demand a hasty release decision, nor restrict nuclear operations to NATO soil. As doctrine-writers struggled to reconcile these competing requirements, they were gradually but irresistibly driven to the same deeper regions of the battlefield on which conventional threat analyses had begun to focus. Thus, the development of nuclear employment doctrine came to reinforce the shift of conventional attention to what became known as "The Extended Battlefield."

The third and last pressure for revision of the 1976 doctrine followed directly from this growing interest in expanding the battle area. For as a practical matter, the Army lacked the resources to do so on its own. If the answer to existing operational weaknesses lay in any significant way in increasing the depth of the battlefield, the Air Force would have to provide the wherewithal—not forever, perhaps, but for a long time to come. An "extended" battle must necessarily be an AirLand battle. What was to become the sobriquet of the Army's new doctrine was thus no mere rhetorical device or political flim-flammery, but rather frank acknowledgment of an operational imperative.

The various pressures just described culminated in August 1982 in the publication of a completely new edition of FM 100-5. The extent of the revision involved is perhaps best illustrated by a formatting decision: although the 1976 edition had been printed looseleaf, to permit page-by-page revision, the 1982 edition was completely reprinted in conventional bound form.

Nevertheless, briefers from the Army's Training and Doctrine Command (TRADOC) insisted that AirLand Battle doctrine was an evolutionary, not a revolutionary change. They were right—but not about the doctrine. It was the Army itself which had evolved and, by 1982, had at last shucked off its own institutional "Vietnam syndrome." With the promulgation of AirLand Battle doctrine, the Army was finally back in the business of winning. The doctrine itself was neither evolutionary nor revolutionary, but reactionary. AirLand Battle doctrine was a "return to basics" with a vengeance:

> [AirLand Battle] doctrine is based on securing or retaining the initiative and exercising it aggressively to defeat the enemy. Destruction of the opposing force is achieved by throwing the enemy off balance with powerful initial blows from unexpected directions and then following up rapidly to prevent his recovery.[7]

Evolutionary? Hardly. These opening lines from FM 100-5's chapter on "Combat Fundamentals" would have surprised neither

Patton nor MacArthur—nor even, for that matter, Hannibal or Frederick the Great. What is perhaps more to the point, they would not have surprised the 1913 French General Staff, whose own revised Field Service Regulations opened with the following rather more flamboyant language: "The French Army, returning to its tradition, henceforth admits no law but the offensive."[8]

The point here is not that FM 100-5 constitutes an American version of *offensive a outrance,* or that its authors were in any sense actuated by the sort of metaphysical faith in *elan vital* which affected their French predecessors. It does not, and they were not. On the contrary, FM 100-5 is an extraordinarily balanced and sensible restatement of tactical principles whose fundamental soundness few soldiers would question. Its basic tenets—initiative, agility, synchronization, and depth—reflect lessons hard learned as far back as Marathon, and rarely successfully challenged on the battlefield.

Instead, the similarity between the two cases lies in what they tell about the respective armies. For in each case, doctrinal revision reflected psychological recovery. In the process, neither the French Army nor the US Army was willing to stomach indefinitely a tactical doctrine which appeared to enshrine the draw as the objective of military operations.

AIRLAND BATTLE: PROBLEMS IN INTERPRETATION

If this underlying impulse is less clear in the present case than in its predecessor, it is only because AirLand Battle doctrine has had to contend with several problems which had no historical antecedents—problems reflecting the diverse concerns which the doctrine sought to amalgamate. While reaction against the Active Defense, recognition of the importance of the deep battle, and renewed concern for nuclear-chemical operations each helped to mold AirLand Battle doctrine, each also gave rise to some fundamental differences in the way the doctrine has since been interpreted. These divergent interpretations were reinforced by prior wide exposure of the "Integrated Battle" and "Extended Battle" precursor concepts; by TRADOC briefings which implied operational prescriptions not contained in FM 100-5 proper; and by the simultaneous emergence of competitive concepts which borrowed the terminology of AirLand Battle doctrine to describe quite different operational concerns.

Tactical Nuclear Employment. The first such issue of interpretation concerned the employment of nuclear weapons. Consonant with the view that such employment must be "integrated" with the conventional battle, references to nuclear operations are scattered throughout FM 100-5. Significantly, the most extensive discussion of nuclear employment principles occurs immediately prior to the discussion of "Deep Battle," and the five target categories identified as "preferred" nuclear targets are all "deep battle" targets.[9]

This peculiarity apart, FM 100-5's various discussions of nuclear employment are individually unexceptional to the point of blandness. Nor are the unique political constraints on such weapons ignored. On the contrary, FM 100-5 is careful to note the special approval requirements and employment limitations affecting both nuclear and chemical weapons.

The problem of interpretation arises not from any particular reference to nuclear weapons, but from their overall treatment. Notwithstanding the occasional *caveats,* the dominant impression conveyed by FM 100-5 is that once released, nuclear weapons are simply another—albeit uniquely potent—form of fire support. And this impression has been further reinforced by TRADOC briefings which heavily emphasize the tactical importance of timely nuclear attack.

There is nothing very novel about this view. Its roots go back at least to 1955, when President Eisenhower himself commented, in regard to tactical nuclear weapons:

> Where these things are used on strictly military targets and for strictly military purposes, I can see no reason why they shouldn't be used as you would a bullet or anything else.[10]

But while it may not be novel, the view that nuclear weapons are simply "bigger bullets" is anathema to NATO. On the contrary, where the Alliance is concerned, even battlefield nuclear employment is a strategic, not simply an operational, matter. And the agreed language defining the conditions in which—and more important, the purposes for which—nuclear employment might be authorized is perhaps the most carefully worded in the entire NATO planning documentation.

It is hardly surprising, then, that even before FM 100-5 was published, the apparent insensitivity of "Integrated Battle" concepts to the prevailing NATO view that any nuclear use is strategic in character and purpose caused major perturbations in Europe. The negative reaction was only magnified by the association of nuclear weapons with emerging "Extended Battle" concepts. The immediate inference drawn from the marriage of a strictly operational approach to nuclear employment with an emphasis on deep attack was that the US Army was seeking to commit NATO to early nuclear use. Indeed, some saw in AirLand Battle doctrine a transparent effort to reverse NATO's long-standing refusal to grant preconditioned release of nuclear weapons. Thus, paradoxically, an emphasis on deeper nuclear targeting originating at least partly in the need to accommodate a lengthy political decision process instead found itself indicted as an attempt to curtail or bypass that process.[11]

The Deep Battle. The second issue of interpretation concerned the function of deep attack itself—whether nuclear or conventional— and its relation to the frontline battle. In principle, this should not have become an issue at all, for FM 100-5 is quite didactic regarding the role of deep attack in the overall operational concept:

> The deep battle component of the AirLand Battle doctrine supports the commander's basic scheme of maneuver by disrupting enemy forces in depth. . . . Deep battle prevents the enemy from massing and creates windows of opportunity *for offensive actions that allow us to defeat him in detail* (Italics added).[12]

In other words, deep attack in AirLand Battle doctrine is subsidiary, not decisive. Its function is to help create conditions permitting the achievement of tactical success. But the mechanism for achieving that success is not deep attack, but offensive maneuver. Indeed, FM 100-5's entire discussion of deep battle consumes fewer than six of the manual's nearly 200 pages. Notwithstanding, this one feature of AirLand Battle doctrine has received more attention than all the other aspects of the doctrine combined, to the point where many now understand deep attack and AirLand Battle doctrine to be synonymous.

As with nuclear employment, this problem stems partly from the conceptual origins of AirLand Battle doctrine. The precursor "Extended Battle" concept was widely briefed in and out of the Army. That concept, as the name implies, focused heavily on deep

51

attack, an emphasis unfortunately carried over into early AirLand Battle briefings.

Overemphasis on the deep battle also reflected its "hi-tech" implications. The deep attack requirements of AirLand Battle doctrine furnished welcome ammunition both to a development community seeking support for advanced target acquisition and attack systems, and to legislators seeking a technological "fix" for NATO's conventional imbalance. Such concerns naturally encouraged the view that deep attack should be considered an autonomous if not decisive battlefield function.[13]

Finally, the latter view was greatly reinforced by concurrent exposure of an independently-developed Supreme Headquarters Allied Powers Europe (SHAPE) concept for conventional deep attack, publicly articulated by the Supreme Allied Commander Europe (SACEUR) in a speech in September 1982, and in several subsequent articles. The origin of this version of deep battle was SHAPE's belief that advanced conventional technologies might be substituted for nuclear weapons, thus raising the nuclear threshold without penalty to deterrence. Although SACEUR carefully—indeed, insistently—distinguished this concept from AirLand Battle doctrine, everyone else routinely confused the two.[14]

Together, these interpretations threatened to turn AirLand Battle doctrine on its head. As deep attack assumed greater prominence, its originally modest objectives and its functional subordination to the ground scheme of maneuver both became obscured, provoking charges on the one hand that deep attack would not work, and on the other, that the Army was reneging on its own operational concept.[15]

Maneuver. But it is the third issue of interpretation which has elicited the most debate, and rightly so, for it is the most serious. It concerns the definition and role of maneuver in AirLand Battle doctrine. "Maneuver," in this context, must be understood as something quite distinct from mere "movement." Both AirLand Battle doctrine and its predecessor depend equally on movement. What distinguishes the two is the purpose of the movement: whereas in the Active Defense, maneuver was fundamentally a defensive process, in AirLand Battle doctrine it is explicitly offensive.

> Maneuver is the dynamic element of combat . . . the employment of force through movement supported by fire to achieve a position of advantage from

which to destroy or to threaten destruction of the enemy. The object of maneuver at the operational level is to focus maximum strength against the enemy's weakest point, *thereby gaining strategic advantage* (Italics added).[16]

The problem arises in the interpretation of the last four words. For the commitment to "gaining strategic advantage" immediately introduces questions of scale. At the extreme, it would imply a commitment to strategic victory.

No such commitment of course appears anywhere in FM 100-5, and for good reason. Even defined in purely operational terms, AirLand Battle doctrine's emphasis on offensive maneuver produced instant reverberations in Europe. Some European critics read in AirLand Battle doctrine's forecast of "non-linear" battles the potential abandonment of the forward defense in favor of mobile operations requiring the exposure of German territory to Soviet penetrations. AirLand Battle doctrine thus inadvertently reopened the question of the viability of forward defense, an issue settled over a decade ago on largely political grounds which the subsequent industrialization and urbanization of Germany have merely solidified. Army protestations that AirLand Battle doctrine implied no such voluntary surrender of territory were not fortified by the concurrent appearance of numerous proposals urging greater attention to "territorial defense."[17]

Even more contentious was the inference drawn by some that AirLand Battle doctrine would require deep cross-border operations by NATO ground forces. For many Europeans, even discussing such operations threatened NATO's image as a purely defensive alliance, and the prospect of German soldiers on East European soil exhumed a host of unpleasant and politically sensitive historical associations.

But the most serious European reactions concerned AirLand Battle doctrine's strategic, not its operational, implications. Europeans added up such terms as "decisive maneuver," "deep attack," and "strategic advantage," and concluded that AirLand Battle doctrine represented nothing less than an indirect assault on both NATO's deterrent strategy and its self-consciously limited war aims should deterrence fail. Their concerns in this regard were only aggravated by TRADOC briefings which appeared to reject reestablishment of the *status quo ante* as a governing military objective; by visuals depicting NATO counteroffensives deep into Eastern Europe; by the continuing chorus of American voices proclaiming the demise of extended nuclear deterrence; and by official and unofficial US discussions of "horizontal escalation" and "war-widening" which appeared to expand AirLand Battle doctrine's operational emphasis on seizing the initiative to global dimensions.

All things considered, it is scarcely surprising that SACEUR flatly disavowed AirLand Battle doctrine. Indeed, the issue became so sensitive that just prior to publication, language was inserted in the introduction to FM 100-5 disclaiming any intention to alter agreed Alliance strategy.[18]

CONVENTIONAL RETALIATORY OFFENSIVE: A CRITIQUE

Since then, Army officials up to and including the former Chief of Staff have been at pains to reassure the Allies that AirLand Battle doctrine connotes no such intention. Unfortunately, collateral evidence on that score is at best mixed. Briefings on AirLand Battle doctrine continue to insist that, should war occur in Central Europe, it will be ended on new terms which "[the Soviets] themselves created." Even more telling is the active interest aroused in official and unofficial circles by a related proposal of Samuel Huntington for a retaliatory offensive strategy (Chapter 2).

Huntington does us the grace of being explicit. In his view, extended nuclear deterrence is no longer credible even against a direct threat to NATO, let alone against Soviet adventures elsewhere. This, and the expanding scope of Soviet global interests and capabilities, markedly increase the likelihood of a Soviet-American war in the 1980s. Both to prevent such a war and to fight it if deterrence fails, the United States and NATO must adopt a conventional retaliatory strategy. And the centerpiece of such a strategy should be a declaratory commitment to a prompt offensive into Eastern Europe with the objective of dismantling the Soviet Union's European empire.[19]

Nothing could be more clear. And here at last, in a doctrinal sense, the American and French cases would converge. For what led France to near-disaster in 1914 was not her tactical obsession with the offensive, costly as that was, but rather her extrapolation of tactical method to strategic doctrine. Describing that process, Tuchman notes:

> Over the years, while French military philosophy had changed, French geography had not. The geographical facts of her frontiers remained what Germany had made them in 1870 .... While French history and development after the turn of the century fixed her mind upon the offensive, her geography still required a strategy of the defensive.[20]

Nevertheless, Tuchman tells us, bedazzled by "the idea with a sword," the French Army ignored both its own strategic circumstances and the abundant evidence of Germany's enveloping intentions, and instead hurled itself forward in a "prompt retaliatory offensive" which cost France over 200,000 dead, the four-year occupation of much of her territory, and, but for the Miracle of Marne, her life. "It was one of those lessons, a survivor realized afterward, 'by which God teaches the law to kings.'"[21]

Unfortunately, it is a lesson which seems to require periodic reteaching: not the risks of the offensive, for these are conditional; but the risk involved in transmuting tactical and operational prescriptions, however valid in their own terms, into strategic injunctions which disregard prevailing circumstances.

The most obvious such circumstance is the continuing conventional imbalance in Europe. According to SACEUR, NATO's conventional resources are already doubtfully adequate to contain a serious Warsaw Pact attack for more than a short period, however the defense were conducted.[22] To suppose that ground combat assets of any size could be spared to undertake the sort of operation proposed by Huntington is simply quixotic. It is one thing to conduct an aggressive maneuver defense of the kind envisioned by FM 100-5, but it is quite another to commit already inadequate ground forces to a precipitate strategic offensive against greatly superior numbers on the optimistic (and improbable) assumption that the threat of such an operation would significantly alter Soviet planning estimates. Certainly no such thing occurred in the 1914 case. On the contrary, German planners both anticipated and hoped for a French offensive through Alsace. And in the view of many historians, it was only Moltke's ill-conceived decision to exploit the failure of that offensive at the expense of Schlieffen's right wing which doomed the latter to defeat. Presumably Soviet planners also read history; it is hard to believe they would make a similar mistake.

Nor should we accept the facile proposition that the mere appearance of western formations on East European soil would precipitate an instant Warsaw Pact mutiny. Whatever the subliminal attitudes of East European publics toward their association with the USSR, their armies have demonstrated quite convincingly their preparedness to maintain this association. It would require a good deal more than a few Western divisions to rupture that commitment—and were the divisions German, the result might well be the opposite.[23]

But the operational objections to the proposal are if anything less significant than the political and strategic objections. The former are the more easily defined, given the furor over AirLand Battle doctrine's far less explicit strategic content. Put simply, NATO does not seek the military destruction of the Soviet empire. Any attempt by the United States to press an offensive conventional strategy on NATO is therefore likely to confirm already widespread European doubts about America's strategic prudence; and, in the process, still further diminish the perception of common security interests essential to continued Alliance cohesion. If we seek to extract ourselves from the European security problem, in short, such an attempt to revive "roll back" is a good way to do it.

But while fatal Alliance dissension would be far and away the most likely result of any effort to turn AirLand Battle doctrine into a new NATO strategy, success in such an effort might be still worse. For the overriding strategic condition still obtains: nuclear weapons will not go away. If it is difficult to believe that the Soviets would be willing to absorb the controlled nuclear escalation upon which NATO's current strategy ultimately depends, it is even less believable that they would tolerate the physical loss of control over their Western buffer while any capability remained unexploited. Thus, the most probable result of a Western commitment to a conventional retaliatory offensive strategy would be an increased risk of the very nuclear war such a strategy is intended to supplant—but given the stakes, one even less likely to be limited. That it would be initiated by the USSR rather than the West would in the end matter little to either.

CONCLUSION

And that brings us to the final issue. For by making explicit the strategic position to which extrapolation of AirLand Battle doctrine ultimately tends, Huntington reminds us of the fundamental tension afflicting Western military policy. For over two decades, NATO strategy has sought to reconcile deterrence with limitation, to make aggression less likely without making war itself more costly. It has never been a happy marriage—least of all for soldiers, whose tolerance of the draw as a strategic objective is only marginally greater than as an operational or tactical one. But it was and remains a necessary marriage—for the continued cohesion of the Alliance, and for the continued domestic support without which no Western military strategy can endure.

AirLand Battle doctrine is the right tactical doctrine for the Army. But, as FM 100-5 itself points out, it was designed to win battles, not wars. AirLand Battle doctrine alters neither our military capabilities nor the strategic objectives and conditions which govern their employment. If we forget that, if instead we delude ourselves that what makes operational sense for divisions and corps necessarily makes strategic sense for nations and alliances, history will not call the result "victory."

ENDNOTES

1. Barbara W. Tuchman, *The Guns of August* (New York: Dell Publishing Co., 1965), p. 48.

2. *Ibid.,* p. 51.

3. Edward L. King, *The Death of the Army: A Pre-Mortem* (New York: Saturday Review Press, 1972).

4. Zeb B. Bradford, Jr., and Frederic J. Brown, *The United States Army in Transition* (Beverly Hills, Calif.: Sage Publications, Inc., 1973), pp. 103ff.

5. US Army War College, *Army Tasks for the Seventies* (Carlisle Barracks, Penna.: US Army War College, June 1972). The writer himself attended the cited meeting with the Secretary of the Army.

6. Among the published critiques, see William S. Lind, "Some Doctrinal Questions for the United States Army," *Military Review,* vol. 57, no. 3 (March 1977), pp. 54-65; Steven L. Canby, "NATO: Reassessing the Conventional Wisdoms," *Survival,* vol. 19, no. 4 (July/August 1977), pp. 164-168; Edward Luttwak, "The American Style of Warfare and the Military Balance," *Survival,* vol. 21, no. 2 (March/April 1979), pp. 57-60; and Wayne Downing, "U.S. Army Operations Doctrine," *Military Review,* vol. 61, no. 1 (January 1981), pp. 64-73. For Army reactions, see John J. Fialka, "Matters of Defense," *Wall Street Journal,* April 13, 1982, p. 52.

7. US Army,FM 100-5, *Operations* (August 20, 1982), p. 2-1.

8. Tuchman, *The Guns of August,* p. 51.

9. FM 100-5, p. 7-12.

10. Samuel P. Huntington, *The Uncommon Defense* (New York: Columbia University Press, 1961), p. 80.

11. Walter Pincus, "Army Would Like Advance Authority to Use A-Weapons," *Washington Post,* July 21, 1982, p. 18; "Atom Plan is Denied by Army," *Chicago Tribune,* July 22, 1982, p. 4.

12. FM 100-5, p. 7-13. See also p. 7-2.

13. Benjamin F. Schemmer, "NATO's New Strategy: Defend Forward, But Strike Deep," *Armed Forces Journal International,* vol. 120, no. 3 (November 1982), pp. 50-68. See also Michael R. Gordon, "E.T. Weapons to Beef Up NATO Forces Raise Technical and Political Doubts," *National Journal,* vol. 15, no. 8 (February 19, 1983), pp. 364-369.

14. Bernard W. Rogers, "The 'Strike Deep' Initiative," *Armed Forces Journal International,* vol. 120, no. 5 (January 1983), p. 5. See also editor's introduction to Roger's more recent article "ACE Attack of Warsaw Pact Follow-On Forces," *Military Technology,* vol. 7, no. 5 (May 1983), pp. 38-50.

15. On the first point, see Trevor N. Dupuy, "Why Strike Deep *Won't* Work," *Armed Forces Journal International,* vol. 120, no. 5 (January 1983), pp. 56-57. On the second, see Gary Hart, "The Army's New Fighting Doctrine," *New York Times Magazine,* January 16, 1983, p. 70.

16. FM 100-5, p. 2-4.

17. See Steven L. Canby, "General Purpose Forces," *International Security Review,* vol. 5, no. 3 (Fall 1980), pp. 317-346; Norbet Hanning, "Can Western Europe Be Defended by Conventional Means?" *International Defense Review,* vol. 12, no. 1 (January 1979), pp. 27-34; and Udo Phillip, "NATO Strategy Under Discussion in Bonn," *International Defense Review,* vol. 13, no. 9 (September 1980), pp. 1367-1371.

18. The key language is in the Preface (p. 1) and in the introduction to Chapter 1.

19. Samuel P. Huntington, ''The Renewal of Strategy,'' in *The Strategic Imperative: New Policies for American Security,* ed. Huntington (Cambridge, Mass.: Ballinger, 1982).

20. Tuchman, *The Guns of August,* p. 51.

21. *Ibid.,* p. 264.

22. Interview with Bernard W. Rogers, *Armed Forces Journal International,* vol. 121, no. 2 (September 1983), pp. 72ff.

23. The interpenetration of Soviet and non-Soviet Warsaw Pact armies at the leadership level makes the argument for massive military defection still weaker. See especially the discussion in Viktor Suvarov's *Inside the Soviet Army* (New York: Macmillan, 1982), pp. 10-21.

CHAPTER 4

STRATEGIC AND DOCTRINAL IMPLICATIONS OF DEEP ATTACK CONCEPTS FOR THE DEFENSE OF CENTRAL EUROPE

by

Boyd Sutton, John R. Landry, Malcolm B. Armstrong, Howell M. Estes, and Wesley K. Clark

NATO's adoption of flexible response in 1967 reflected the growing awareness on both sides of the Atlantic that a deterrent strategy based solely on the threat of mutual annihilation was neither credible nor practical in the face of growing Soviet strategic and theater nuclear capabilities. Deterrence required strong conventional forces capable of responding to conventional attack as well as the threat of escalation to nuclear war.

Alliance nations committed themselves to a series of ambitious defense programs to improve conventional capabilities and, while execution of these programs has been neither uniform nor complete, they yielded a remarkable improvement in NATO's conventional forward defenses. The relentless Soviet buildup of its military forces, both nuclear and conventional, however, has consistently outpaced Alliance efforts. By the end of the 1970s, many analysts questioned whether continued dedication to existing defense programs was likely to produce a conventional defense equal to the needs of flexible response. At the same time, few suggested that increased reliance on nuclear weapons made any sense, particularly in view of substantial Soviet strategic and theater nuclear force gains.

This dilemma predictably has produced two responses: to change the basic strategy or to seek substantial improvements in conventional force capability to make the existing strategy work. Samuel P. Huntington's concept of a "conventional retaliation" strategy is one of several proposals designed to change the strategy. Conventional "deep attack" concepts, and the emerging technologies that make such concepts feasible, provide the affordable potential to improve conventional capabilities and restore credibility to the current strategy. This chapter examines two concepts—Supreme Headquarters Allied Powers Europe's (SHAPE) Follow-On Force Attack (FOFA), and the US Army's AirLand Battle (ALB) doctrine—that rely heavily on deep attack for their success.

NATO DEFENSE AND DEEP ATTACK CONCEPTS

Deep attack concepts emerged from the interaction of three related but distinctly different influences: NATO concerns with the unrelenting Warsaw Pact conventional force buildup in Central Europe since the late 1960s; dissatisfaction with what was widely regarded by the US Army to be an excessively reactive defense doctrine; and the development of emerging technologies that offer the potential for substantially better target acquisition and conventional weapons lethality.

The outline of Soviet operational strategy in Central Europe has been clear for at least a decade. That strategy seeks a quick penetration of NATO's forward defenses and a rapid advance into the strategic depths of Alliance territory to forestall full mobilization and reinforcement from the United States and to bring about early military and political collapse. On the ground, this strategy is pursued by concentrating overwhelming force and by committing forces in succeeding echelons to maintain the momentum of combat operations. First echelon forces fix NATO's forward forces in position or destroy them; second echelon forces complete their destruction and flow through lines of least resistance to penetrate deeply into NATO's rear to disrupt the Alliance's ability to reinforce. Warsaw Pact follow-on forces are thus critical to the overall strategy.

By the mid-1970s, this strategy and Soviet improvements in conventional doctrine, armaments, and force structure converged to create severe doubts whether NATO could mount a credible direct defense. Two aspects of the Soviet buildup were particularly threatening. First, the Soviets appeared to be aiming for a short-warning attack capability by reducing their reliance on early reinforcement from the Western Military Districts of the Soviet Union. Larger divisions, armed with greater numbers of more advanced weapons, improved logistics support, and increased readiness were all part of this effort. These improvements threatened to reduce substantially both strategic and tactical warning, and thus reduce NATO's ability to countermobilize, deploy forces forward, and prepare defensive positions.

Second, the Soviets launched an ambitious effort to achieve and maintain rapid attack advance rates. Soviet armaments concentrated on systems to exploit a high speed attack including infantry combat vehicles, mechanized artillery, increased tactical aviation and helicopter gunships to provide continuous close support, and more effective antiaircraft weapons to suppress enemy air. Additionally, emphasis was placed on developing more rapid and flexible command and control procedures for ground, air, and air defense forces. Specific efforts focused on an improved capability to commit follow-on forces where and when they would achieve the greatest effect, to shift combat effort from one axis to another in accordance with rapidly changing circumstances, and to coordinate the efforts of adjacent combat formations.[1]

Together, these trends attacked the credibility of NATO's flexible response strategy by threatening to force an early political collapse of the Alliance before NATO could make the decision to escalate to the use of nuclear weapons. Without a strong direct defense capable of providing time for Alliance leaders to reach a nuclear decision and confidence that continued resistance is militarily and politically feasible, the very basis of flexible response is seriously eroded.

While US Army commanders in Europe were concerned with these trends, they were also disturbed by the Army's existing land force doctrine—the Active Defense—which many believed to be inappropriate to the nature of the Soviet challenge. The doctrine of Active Defense as it is sometimes called, was promulgated in FM 100-5, *Operations,* in 1976. It sought to concentrate firepower on enemy units in "killing zones" and then to defeat these units sequentially as they arrived in the main battle area. Active Defense was criticized on three counts. First, it was regarded as excessively reactive, ceding the initiative to the attacker by discouraging maneuver of forces against enemy vulnerabilities. Second, it was said to focus inordinately on massing firepower at the point of an attempted Soviet breakthrough, where the enemy's strength was concentrated. This tactic was thought to be flawed, both because lateral movement of friendly forces in the face of Warsaw Pact massed formations is problematical and because combat force ratios in these areas continuously and overwhelmingly favored the attacker. Finally, by seemingly endorsing an attrition oriented defense, the doctrine was assumed to risk early exhaustion of forward forces by subjecting them to continuous operations against fresh, fully structured enemy units fed into the forward battle from reinforcing echelons. In short, Army commanders doubted success of a doctrine which confronted wave after wave of echeloned Warsaw Pact forces with a linear, positional defense.

As Army commanders wrestled with this problem, technological developments converged to offer some promise of offsetting Soviet quantitative superiority. Developments in a number of areas, particularly sensors, data processing, terminal guidance, and conventional munitions effects offered a potential for more effective engagement of mobile targets at extended ranges.

As the potential of these technologies became more apparent, proposals for their strategic and tactical application proliferated. Some spoke of strategically isolating the forward battlefield by destroying an enemy's capability to reinforce, while others envisioned attacks against key Warsaw Pact command and control capabilities to deny the aggressor an ability to employ effectively his echeloned formations. Virtually all proposals suggested that emerging capabilities would reduce—and some even said eliminate—NATO's reliance on nuclear weapons by stiffening the forward defense and by providing conventional arms with a lethality approaching that of low yield nuclear weapons.[2]

By the late 1970s, these three factors converged to produce doctrinal change which incorporated emerging deep attack technologies. The key assumption at the foundation of all deep attack concepts is that NATO and Warsaw Pact forces *in the first echelon* are relatively evenly matched and that NATO can provide a credible forward defense at the conventional level *if* Warsaw Pact reinforcing echelons can be kept out of the forward battle or at least allowed into it when and where most advantageous to NATO. This essential equivalence in the first echelon arises because NATO's forward defense policy requires virtually all of its ground combat power to be placed forward in the "first echelon." On the other hand, while the Warsaw Pact has a theater-wide numerical superiority in units and weapons, terrain limitations, the requirement to operate somewhat dispersed because of the potential for use of nuclear weapons, and their basic doctrine of operating from deeply echeloned formations virtually preclude the Warsaw Pact from using its theater-wide advantage at the line of contact. Using the inherent advantages of defense, and so long as the Warsaw Pact's large follow-on forces can be held out of the forward battle, NATO's chances of executing a successful forward defense are improved.

AirLand Battle Doctrine. AirLand Battle doctrine focuses on combat operations of the Corps and subordinate elements; it seeks to win battles, not wars, recognizing that successful military operations are an indispensable but insufficient guarantee of victory.[3] The doctrine seeks to employ existing weapons and forces more effectively rather than prescribe the development of new ones.[4] Thus, while recognizing that it would benefit from technologies expected to be fielded in the next several years, it is not predicated on their availability.

Within these limits, the purpose of AirLand Battle doctrine is to deny success to an aggressor's attack by seizing and maintaining the initiative.[5] To achieve this objective, the doctrine specifies the need to break the momentum of the enemy's attack, to destroy synchronization among the elements of attacking forces, and to defeat those forces piecemeal. Two battles must be fought simultaneously and in close coordination: a forward battle against committed units, and a deep battle against uncommitted forces both to delay and disrupt their commitment to the forward battle and to create opportunities for subsequent maneuver against them. While the doctrine seeks a balance between maneuver and firepower, maneuver is particularly emphasized as the means of concentrating strength against enemy weaknesses and gaining a position of operational advantage from which to mass effective fires against enemy vulnerabilities.[6]

At the outset, AirLand Battle recognizes that deep attack—or the deep battle, as it is termed—is a prerequisite to successful execution of the doctrine. Because the doctrine focuses on Corps operations, it envisions the conduct of the deep battle out to 150 to 200 kilometers (km)—the limit of the Corps commander's area of influence. The means to execute the deep battle include all capabilities available to the ground commander, such as artillery, electronic warfare and deception, the maneuver of friendly ground forces, as well as air support. The Army recognizes that "battlefield air interdiction"—a NATO term not yet accepted into US joint doctrine—is the primary means of fighting the deep battle at extended ranges.[7]

The SHAPE Follow-On Force Attack Concept. Shortly after assuming responsibilities as Supreme Allied Commander Europe (SACEUR) in 1979, General Bernard W. Rogers tasked the SHAPE staff to develop a concept for holding at risk Soviet follow-on forces with deep conventional fires. The SHAPE concept seeks to locate and track Warsaw Pact forces during their entire process of deployment, from garrison to battlefield commitment, and to attack them when and where they are most vulnerable. The concept aims at exploiting particularly critical enemy vulnerabilities in the reinforcement process, the rigidity of his planning for an echeloned offense, the density of forces along limited attack routes, and critical transportation facilities.

Like the AirLand Battle doctrine, SHAPE's Follow-On Force Attack concept seeks to link the deep battle with the combat operations of forces in contact, but the means of doing so are vastly different. While AirLand Battle seeks to synchronize the deep battle with the ground commander's scheme of maneuver, the SHAPE concept focuses simply on the centralized application of all deep attack assets to separate first echelon and second echelon forces in order to maintain NATO/Warsaw Pact combat force ratios in the first echelon at a manageable level.

DEEP ATTACK CONCEPTS AND
NATO'S FLEXIBLE RESPONSE STRATEGY

Alliance deliberations regarding the political acceptability and military utility of deep attack concepts will turn first and foremost on their implications for NATO's strategy. To some extent this is true because various US authorities have emphasized their strategic value, but more importantly because Europeans are concerned that new doctrinal designs indicate possible shifts in the agreed strategy of flexible response. Generally, Europeans have been wary of deep attack conceptions, suspecting the worst when any consideration is given to modifying NATO military strategy. These suspicions threaten to derail valuable adjustments to NATO's tactical doctrines—adjustments which could make more credible NATO's flexible response strategy—and to slow the exploitation of promising new conventional technologies.

European Concerns. At the outset, it is important to recognize that some influential segments of the European NATO community have given support to deep attack concepts as potentially productive initiatives to strengthen the Alliance's conventional capabilities. Strong endorsement has been provided, for example, by the British Conservative Party's Defense Committee Vice Chairman, Sir Julian Critchley. In an article written for the *Daily Telegraph* in November 1982, Critchley gave ringing support for both Senator Sam Nunn's call for NATO to adopt AirLand Battle concepts and for General Rogers' outline of the need for a deep attack capability.[8] Similarly, Manfred Woerner, the defense minister for the Federal Republic of Germany (FRG), has urged the exploitation of emerging technologies[9] and the Inspector General of the Bundeswher, General Meinhard Glanz, has stated explicitly the need to prevent the reinforcement of Soviet follow-on forces to assure a strong forward defense.[10] Despite these endorsements, however, major concerns remain.

The principal European concern is a suspicion that the new doctrinal concepts are designed either to provide the basis for a conventional deterrent capability independent of nuclear escalation or to lay the foundation for a warfighting capability in which the use of nuclear weapons is confined to the European theater. The French explicitly communicated the first concern in response to a September 30, 1982 speech that SACEUR gave in Brussels. In that speech, General Rogers noted the significant advantages to be gained by exploiting technology that attack Warsaw Pact follow-on forces; observed that NATO's technological edge permitted development of such a capability; and commented that a four percent gross national product (GNP) allocation to defense by each NATO nation would suffice to procure it.[11] The French newspaper *Le Monde* charged that the concept constituted a move toward a "no early use" policy for nuclear weapons—believed to be no better than a "no first use" pledge—and decried the apparent shift in American perceptions of the role of nuclear weapons from deterring destruction rather than invasion.[12] *Le Monde* also questioned how firmly the United States would continue to support intermediate-range nuclear force (INF) deployments if longer range nuclear systems were to become secondary to conventional deep attack capabilities.[13]

While the French have been the most vocal and explicit in their critique of deep attack concepts—as indeed they have been against any defense initiative which appears to erode the credibility of Alliance nuclear deterrence—other European nations share the concerns from which those criticisms arise. On the one hand, many Europeans, particularly the West Germans, share the French concern that one of the greatest dangers facing the Alliance today is the decoupling of US strategic nuclear guarantees from the defense of Europe. While most Europeans—the French apart—recognize the need for improved conventional forces to make flexible response credible, many suspect that the US intention in pushing hard for such improvements is driven less by a desire to strengthen NATO's threat to escalate a conflict if necessary, than to avoid a nuclear decision altogether. Those who hold this view, point to the "no first use" advocacy of former US officials as confirmation.[14] Europeans fear this development because it would eliminate the incalculability of risks confronting the Soviet Union and thus lower the threshold of conventional aggression.

It should come as no surprise, then, that Europeans would be distressed when various deep attack advocates suggest that emerging conventional technologies "... may provide a conventional military power so formidable as to rival in the tactical arena the deterrent effect nuclear weapons have had on strategic war."[15] Such views focus on what many Europeans believe is a dangerous concept, i.e., a successful conventional option by itself is capable of deterring Soviet aggression.

On the other hand, some Europeans believe that the new US concepts of deep attack may call for both nuclear and conventional capability, thereby leading to plans for the use of nuclear weapons in a warfighting role. They find such a strategy even more frightening than the drive for a conventional-only option. During the Enhanced Radiation Weapon debate in Europe, Manfred Woerner, then a leading West German CDU/CSU defense spokesman, made clear the rejection of a theater nuclear warfighting doctrine confined to Central Europe.

> It is equally clear that a separation of tactical nuclear weapons from the strategic nuclear level is absolutely unacceptable for us Europeans.

> The territory of the USSR cannot be allowed, in theory or practice, to become a sanctuary in the nuclear phase of a conflict in Europe. The Soviet Union cannot be invited to contemplate a war limited to Western Europe, or even to German territory. Moscow must at all times be forced to reckon with the full ladder of escalation.[16]

Some Europeans have noted with concern several briefings of AirLand Battle which have reportedly contained discussions of nuclear targeting decisions which could have been interpreted to imply the need for early, and perhaps even preconditional, release authority for NATO commanders to use nuclear weapons.[17] And, of course, since AirLand Battle doctrine remains focused at Corps operations or below, discussions of nuclear use have generally been confined to their warfighting implications, an understandable fact, but one of disturbing implications for Europeans nonetheless.

Some NATO nations, particularly the FRG, have also been sensitive to assertions that AirLand Battle doctrine envisions the conduct of strategically offensive ground operations. The Bundeswher's Inspector General, while noting the West German acceptance of deep attack concepts in principle, has been extremely careful to reject any notion that German forces are, or will be, structured logistically to attack deep into Warsaw Pact territory.[18]

In part, rejection of such capabilities reflects NATO's defensive orientation. But it also acknowledges that the specter of an offensively capable Bundeswher is acceptable neither to the Soviets nor to the FRG's continental allies.

Finally, suggestions have surfaced from time to time in the European media that the maneuver oriented AirLand Battle doctrine constitutes a retreat from the Alliance's concept of a forward defense. These concerns, erroneous and inconsistent though they might be, reflect considerable confusion regarding strategic vice tactical purposes of deep attack concepts and the doctrinal context to which they are related. This confusion highlights the need for a militarily sound and politically acceptable justification for them if the United States is to continue the considerable progress attained to date.

A Strategic Rationale for Deep Attack Concepts. The litmus test to determine the political acceptability of deep attack concepts is the degree to which they support and reinforce flexible response. For the Europeans, doctrinal conformance with the Alliance's agreed strategy is a measure of continued US commitment to the defense of Europe and our willingness to share the risks as well as the benefits of Alliance membership.

To be consistent with flexible response, deep attack concepts must be clearly portrayed and understood as efforts to provide for a strong direct defense. As such, their purposes must be expressed as efforts to defeat less than full-scale conventional attacks and to deter massive conventional aggression by improving the credibility of Alliance threats of deliberate escalation. This suggests that deep attack concepts must satisfy, at a minimum, three conditions. First, they must contribute to an effective forward defense of Alliance territory so that Pact forces are unable to force a political collapse of NATO either by a limited attack or by seizing significant territory early in a large scale conflict. Implicit in this standard is that deep attack must provide time for Alliance authorities to be able to reach and execute a deliberate nuclear decision. Reaching that decision, should it be necessary, will be anguishing for all NATO nations and each will seek to assure that all alternatives short of escalation have been exhausted. Without time afforded by strong conventional forces, there is some doubt that such a decision could be reached at all in a timely fashion and the credibility of flexible response would be correspondingly eroded. Second, conventional deep attack must not conflict significantly with the execution systems, procedures, or perceptions concerning NATO's means and capabilities to undertake deliberate nuclear escalation. Third, deep attack must not come to be viewed as an alternative to nuclear escalation, either in war plans or in forces developed for their execution.

As concepts for deep attack continue to unfold, it is clear that they can, but not necessarily that they will, meet these criteria. Surely all advocates intend that their concepts strengthen NATO's conventional direct defense, although the precise contribution of any deep attack concept to this purpose will depend not only on funding levels, but also on the situation and conditions existing at the outbreak of hostilities.

The execution of deep attack options in a period of hostilities does not need to conflict with NATO's means and capabilities to undertake deliberate escalation. Separate attack systems can be developed, different control procedures can be employed, and differing targets struck. In this regard, however, it must be noted that efforts to portray improved conventional munitions as having effects equivalent to those of low yield nuclear weapons constitute a two-edged sword, arguing on the one hand for the feasibility of conventional deep attack and on the other suggesting its availability as a substitute for nuclear weapons. Careful examination of these new weapons' characteristics suggests that they can substitute for only the very lowest yield systems, the systems which for reasons of collateral damage restraint would tend to be employed relatively near friendly positions. However, they cannot substitute for the effect of the medium size nuclear weapons which might be contemplated for use against deeper enemy targets.

This suggests that conventional deep attack concepts, properly understood and articulated, cannot stand as an independent strategy. The deterrent effects associated with the risk of uncontrolled escalation to nuclear weapons would be absent. Because conventional weapons, no matter how destructive, do not convey the perception of escalation to strategic nuclear warfare, they cannot substitute for the deterrent effect of tactical nuclear weapons.

STRATEGIC PITFALLS

Threshold Arguments. There remain several pitfalls in developing and articulating the rationale for deep attack. First, what is the effect of a deep attack policy on the nuclear threshold? To suggest that deep attack will raise the nuclear threshold is to imply to some that the ultimate aim is a conventional defense. To focus on the other alternative—a lowered nuclear threshold—seems to suggest some sort of nuclear warfighting, which also raises concerns. The purpose of NATO's strategy is to deter aggression, not simply to prevent the conflict's escalation to the nuclear level. To the extent that improvements in NATO's defenses contribute to the former, they will most effectively assure the latter; and it is in precisely this perspective that deep attack must be viewed and addressed.

Competing Priorities. A second pitfall associated with emerging deep attack concepts is that they complicate the establishment of priorities. Even while NATO's capacity to defend against a short-warning attack erodes, and while the Alliance-Warsaw Pact balance in the forward echelons of the battlefield become more disparate, some deep attack advocates urge the Alliance to devote significant resources toward the development of a technological capability to attack very deeply into Warsaw Pact territory. If economic constraints did not exist, those capabilities, in combination with greater capabilities in the first echelon, would be most welcome. But constraints do exist, and the Alliance must ask where it could most profitably apply marginal increments in resources.

Strategically, the first priority must be accorded to the most dangerous threat to NATO. Currently, and for the foreseeable future, that threat is the Warsaw Pact's ability to achieve a quick breakthrough of NATO's forward defenses. The development of these capabilities has received priority in the Pact's defense improvements and is believed to be a prerequisite to Soviet operational strategy in Central Europe. For that reason, Alliance initiatives must be directed first toward denying such a capability.

The Alliance must direct its priority efforts toward blunting the attack of the first and second tactical echelons of Warsaw Pact forces and those operational follow-on echelons close to the forward battle rather than to strategic follow-on echelons moving forward from the USSR. These forward forces, already located in Eastern Europe during peacetime, are the ones to be committed in the very early days of a conflict and those which pose the most immediate danger of a breakthrough. It would be sheer folly to focus too much on a more distant threat when one closer at hand theatens disaster. Great care must be taken not to incur a strategic deficit by building for the desirable at the expense of the essential.

Nuclear-Conventional Substitutability Arguments. A third, and at least equally dangerous strategic pitfall to the current evolution of deep attack concepts, is the tendency by some to speak as if emerging technologies offer an opportunity to substitute deep conventional capabilities for theater nuclear forces. Technologists in particular favor such an approach by noting the growing accuracy of delivery systems and the increased lethality and area coverage of conventional munitions. Moreover, disenchantment is growing among many defense analysts with the utility and survivability of battlefield nuclear weapons and increasing insistence that they be eliminated wholly from Alliance inventories. Together, these trends, whether intended or not, appear to argue for some substitution of conventional munitions for theater nuclear weapons.

Unfortunately, such a change fails to address the broader strategic requirements of flexible response. Despite the fact that conventional munitions have and will continue to become more militarily effective against an array of targets that were formerly vulnerable only to nuclear targeting, the fact is, theater nuclear weapons have always sought more than military effects on the battlefield. They have sought to achieve political results by threatening escalation of the conflict making the costs and risks of continued aggression clearly disproportionate to perceived gains. No matter how militarily effective conventional munitions may become—and they are not yet nearly so effective as nuclear weapons—they simply cannot convey those risks.

On the other hand, the utility of battlefield nuclear systems is clearly suspect, because they simply are not a credible means of threatening a NATO "first use." First, given the tactical circumstances necessary to provoke an Alliance resort to nuclear weapons, battlefield systems are likely to prove more devastating to Alliance territory and values than to those of the Warsaw Pact. The FRG, in particular, has been adamant that the destructive consequences of nuclear escalation must not be confined to NATO's territory and for that reason they have supported an improvement in longer range nuclear systems. Second, given the necessity for continuing political control of nuclear weapons, in war as well as peace, and the restrictive release procedures maintained for that purpose, there has been increasing doubt in many quarters whether battlefield systems are sufficiently responsive to pose a threat to the mobile target arrays against which they are aimed. Finally, NATO's political authorities have long recognized that in many circumstances the use of battlefield nuclear weapons alone may not attain either the military or political ends sought in an Alliance "first use" of nuclear weapons.

It must be recognized, however, that despite these inherent difficulties, battlefield nuclear systems have performed a critical function by forcing Warsaw Pact forces to operate dispersed in a "nuclear scared" posture. Despite the promised development of drastically improved conventional munitions, nuclear systems still offer greater damage capabilities. Without the threat of short-range battlefield systems, opposing forces would be able to concentrate with far less vulnerability, thus aggravating NATO's conventional defense problem in the main battle area. Moreover, these tactical nuclear systems serve a vital role in deterring Soviet use of its expanding arsenal of short-range nuclear systems.

An analysis of flexible response and deep attack options leads to two major policy conclusions. First, NATO's conventional and nuclear capabilities are not separate entities, but synergistic components of an effective defense posture. One cannot replace the other and major deficiencies in one cannot be compensated for by improvements in the other. For that reason, it is essential that conventional deep attack operations be recognized as complementary rather than as an alternative to proposed modification to nuclear options which rely on similar emerging weapons developments. One such option for the initial use of NATO nuclear weapons is to direct NATO's longer range nuclear systems toward the disruption and destruction of Soviet operational and strategic follow-on echelons before they are committed to the forward battle. The political utility of such an option is that it clearly would convey NATO's intent to deny Moscow its war aims by threatening more dangerous escalation should hostilities continue. Further, such an option meets the demanding requirements expected of a NATO "first use": it communicates the seriousness of NATO's purpose; it signals a willingness to escalate further; and it insures that nuclear options are militarily responsive. Finally, given Alliance deployment of Pershing II and Ground Launched Cruise Missiles (GLCM) and completion of programmed improvements to Command, Control, Communications, and Intelligence facilities (C^3I), such a capability is achievable in the relatively near term.

The second implication is that NATO cannot afford to dispense altogether with a capability to target the forward attacking echelons of Warsaw Pact forces with nuclear means. This does not mean that obsolescent weapons could not be removed from Alliance inventories, as NATO recently announced that it would do, or that artillery fired atomic projectiles are necessarily the most effective means of accomplishing such objectives. It does mean, however, that NATO must retain a full spectrum of nuclear capabilities to "hold at risk" the entire depth of attacking forces in order to assure their dispersal and to deter the use of Pact nuclear systems at all levels.

DOCTRINE

The two specific deep attack concepts under discussion in the Alliance, the US Army's AirLand Battle doctrine and the SHAPE Follow-On Force Attack concept, are similar in several respects: both recognize the significance of the Soviet forces echeloned in depth; both recognize the importance of seeing and attacking in depth; both recognize that airpower is critical for this purpose; and both recognize that air interdiction must be more closely orchestrated to affect the ground battle. Despite these similarities, however, the two approaches are surprisingly different both in their underlying assumptions and in their implications.

Fundamental Distinctions. The two approaches reflect their differing origins, the differing concerns of their originators, and the differing parameters within which they were developed. AirLand Battle was developed by the US Army, with Air Force participation, for Corps level and below to enable US forces to defeat a technologically equivalent, numerically superior opponent on any battlefield in the world. Implicit in AirLand Battle doctrine is concern that, to defeat even first echelon opposing forces, US forces must alter their approach to warfare by stressing maneuver and fighting in depth. The SHAPE concept, on the other hand, was prepared by an integrated Alliance headquarters to deal solely with the problem of the opposing forces' reinforcing echelons theater-wide, recognizing that Corps will fight according to their various national doctrines.

Central to AirLand Battle doctrine is the recognition that commanders must be able to detect and delay or disrupt those opposing forces echeloned in depth which could interfere with operations against the enemy's first echelon. The doctrine requires that a Corps strive to maintain surveillance of an area of interest large enough to give ninety-six hours notice of approaching significant enemy forces and must have the ability to influence those opposing forces up to seventy-two hours away from the main battle.[19] With these capabilities, the Corps commander is to plan and execute battle actions enabling him to wrest the initiative from the attacking enemy force. As this planning-execution window is compressed, the difficulties and risks associated with seizing the initiative will increase accordingly. In the European theater, this seventy-two hour window, plotted on a map as the Corps area of influence, could extend as much as 150-200 kilometers forward of the main battle.[20]

Some of the key procedural problems for implementing AirLand Battle in NATO are: allocation of battlefield interdiction sorties down to Corps and below; recognition of an area of influence beyond the Fire Support Coordination Line in which sorties would be available to supplement organic Corps deep attack systems; and early sortie allocation to enable airpower to be integrated fully into the planning of the ground force.

Current procedures allow for the first, are beginning to take cognizance of the second, but have made little progress in the third area. Indeed, this notion of early commitment of significant air interdiction forces, which runs counter to the notion of centralized control of air and its dispatch *en masse* to the most crucial portion of the theater, poses a key difficulty in the full implementation of AirLand Battle as it was originally conceived.

Offensive air support, including close air support, battlefield air interdiction, and reconnaissance can be allocated down to Corps level. The ground commander's area of responsibility forward of the close battle is normally terminated at the Fire Support

Coordination Line, approximately twenty-five kilometers forward; beyond this distance targets are the responsibility of the air component commander, though some measures are now in process which will assure that to depths of seventy to one hundred kilometers beyond the front line of engaged forces, air interdiction of targets will be coordinated with the ground commander. Procedural requirements for planning centralized daily apportionment and allocation of air resources at the air component command and subordinate air operations centers are time consuming. The long-range lead time planning procedures for allocating the critical battlefield air interdiction sorties have not been developed. This means that commanders at Corps level and below have not been able to develop plans which adequately integrate the principal means of deep attack—airpower. Instead, they have been able to do little more than nominate in priority their requests for air reconnaissance and attack targets, without knowing until a few hours prior to the battle if they will get the support.

The SHAPE concept, on the other hand, poses no requirements for developing new allocation procedures; it was designed to take advantage of the very centralized air allocation procedures which cause difficulty for AirLand Battle.

Thus, there is this disjuncture between the AirLand Battle and the SHAPE concept: AirLand Battle thrives on the early allocation of airpower to support the ground commander, a process which reduces the extent of centralized control and application; the SHAPE concept, however, plans for more traditional use of airpower through centralized air allocation and application theater-wide. This disjuncture has profound implications for how the ground war can be fought.

Contrasting Implications. AirLand Battle posits maneuver by up to division sized forces to seize the initiative and defeat forward enemy forces. To do this, however, some assurance must be provided to the Corps that the enemy's echeloned follow-on forces will be prevented from interfering with that maneuver. Corps commanders need sufficient time to allow friendly forces a reasonable prospect of defeating the enemy first echelon. Without some increased confidence that the required air assets will be available to support the ground battle, it will be even more difficult, and risky for the Corps to attempt to assume the initiative. In this context, then, the disjuncture in the allocation and application of airpower between the Army's AirLand Battle and the SHAPE concept reflects the tension between a doctrine which recognizes a need to maneuver and one which takes a firepower-attrition approach to the battlefield. Alternatively, this disjuncture may be viewed as the conflict between the warfighting concerns necessarily prevalent at Corps level and below, where the foremost consideration must be to prevail in battle against attacking Pact forces, and the deterrent focus characteristic of higher commanders, where the Soviets must be "persuaded" to call off the attack either because they cannot "win" or because they cannot afford the costs that winning would entail.

Moreover, the Army's Corps-level doctrine is based on the conclusion that maneuver is essential to defeat the Pact's first tactical echelon, and this, in turn, will require greater synchronization of air interdiction with ground forces than is envisioned in current procedures. From the theater level, however, air interdiction necessarily constitutes the primary means short of nuclear exchange to attack the Pact rear. It constitutes an essential element of intrawar deterrence whose contribution exceeds any materiel damage inflicted. While this deterrent perspective does not rule out greater synchronization of air and ground forces, it does serve to constrain the amount of air resources allocated to the support of ground forces, as well as to argue for the preservation of the centralized level at which these forces are allocated so that airpower can be massed more easily to support theater requirements and priorities.

A third perspective on these two deep attack concepts is apparent in the time each requires to have significant military (*vice* political) effect. In this context, the AirLand Battle implies a near term impact through the synchronization of air and ground forces against relatively close opposing formations. The SHAPE concept, though currently limited by the combat radii of most NATO attack aircraft, attempts to impede reinforcement of the forward area and affect the war rather than the immediate battlefield. In a time of acute crisis, however, more immediate military needs would tend to receive priority. Faced with the choice between interdiction of opposing forces still more than three days from the battle or staving off a breakthrough in the Central Region, the commander would presumably opt to trade future security for present survival (see Figure 4-1). While this consideration does not negate the

FIGURE 4-1
COMPARISON OF AIR INTERDICTION ZONES

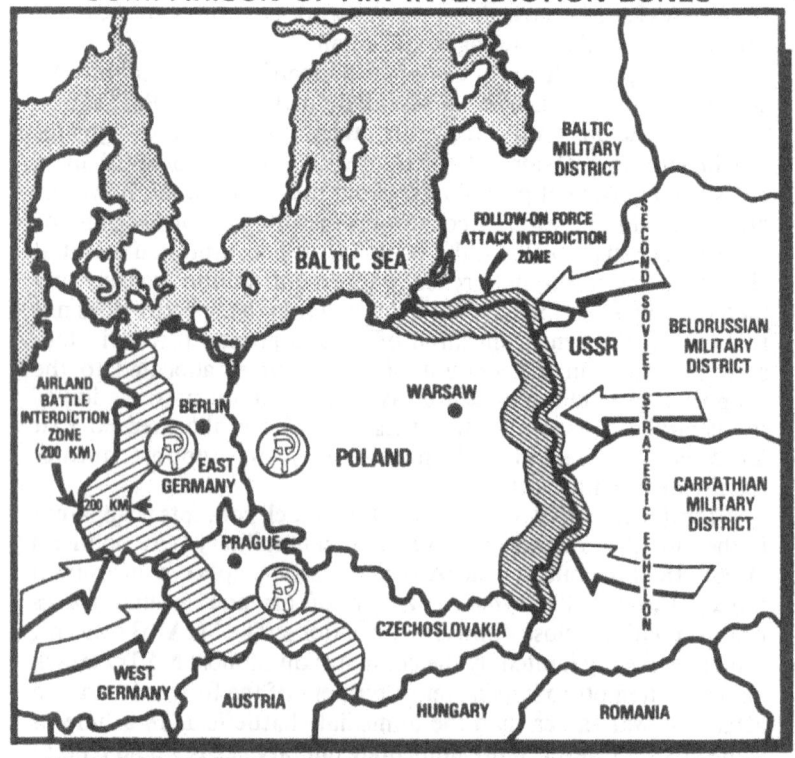

NOTE: CURRENTLY AIRLAND BATTLE DOCTRINE APPLIES ONLY TO US FORCES. MAP SHOWS EXTENTION OF
AIRLAND BATTLE INTERDICTION ZONE TO OTHER ALLIED FORCES.

desirability of possessing the capability to strike very deep, it argues for the necessity of placing first priority on developing the most effective procedures and systems for applying airpower to influence the close battle.

Examination of the preconflict deterrent implications of the two concepts reinforces the foregoing consideration. To be sure, virtually all improvements in NATO capabilities will tend to strengthen the overall deterrent, but the two concepts impact somewhat differently on what might be called the crisis stability of the deterrent. Insofar as the capability for deep interdiction is improved, the risks of a Warsaw Pact "bolt from the blue" or limited mobilization attack will be increased, thus, presumably reducing Pact incentives to attack. But such reductions in Warsaw Pact incentives may not be as strong in the case of a full mobilization attack, since the Pact would be able to bring forward its major formations prior to the onset of hostilities, negating the full impact of a deep attack concept. AirLand Battle, on the other hand, would be significant against either a limited or a full mobilization attack, since AirLand Battle threatens the destruction of the opposing force echelon in contact.

Finally, it must be noted that since AirLand Battle doctrine was not designed specifically for the NATO context of separate national Corps, therefore, its implementation within the Central Region poses a unique set of problems. These problems are distinct from those of implementing the SHAPE concept. For example, if other nations do not employ the US area of influence concept forward of the main battle, then the development of deep intelligence may be asymmetrical despite the best efforts of inter-Allied intelligence sharing. Also, there is danger that the air allocation system may find itself in the dilemma of either taking special cognizance of US Corps interdiction requests—conventionally viewed as a suboptimization of airpower—or frustrating the very air procedures most likely to employ airpower to its fullest effect. Together these considerations suggest that AirLand Battle may constitute a form of doctrinal encroachment on the procedures of the other national Corps unlike that heretofore experienced.

RECONCILING TWO DISTINCT APPROACHES

Is there a necessity to choose between these distinct approaches? Should both proceed independently? Can they in some way be combined? The AirLand Battle and SHAPE concepts would seem to have both complementary and competitive aspects. Certainly their differing implications with regard to warfighting versus deterrence, the time periods required for military impact during conflict, and preconflict deterrence would appear in the main complementary, though the actual wartime allocation of resources for various purposes might be competitive. On the other hand, the difference in the implied locus of control of air assets implies alternative approaches to warfighting which are definitely competitive. Also, the tendency of the AirLand Battle doctrine to infringe on other national Corps-level doctrines requires careful consideration.

There are three obvious standards by which these complementary and competitive aspects may be examined. First, can the two different approaches be made procedurally compatible, so that the theater commander may employ either or both simultaneously? Second, can AirLand Battle be fitted into the NATO national Corps approach as one nation's battle doctrine? Third, is the military hardware to implement these approaches similar and compatible, or will the theater commander's flexibility in wartime be curtailed by procurements underway today?

Procedures. At present the centralized daily air allocation procedures clearly restrict the employment of AirLand Battle in Europe. Fortunately, there appear to be efforts underway in US circles to deal in part with the air allocation issue. A modification of existing air allocation procedures entitled Joint Air Attack of the Second Echelon was coordinated between Army and Air Force authorities in December 1982. Under these procedures, a Battle Coordination Element would operate at the Tactical Air Control Center for the purpose of prioritizing the Army air interdiction requests and insuring Air Force appreciation of ground maneuver requirements.

This modification of the Air Support Operations Center (ASOC) concept and the recent Air Force decision to endorse AirLand Battle doctrine represent very positive steps. It remains to be seen, however, if these agreements can be transformed into operating procedures within the theater which demonstrate their adequacy and their timeliness. In particular, various types of *early* allocation systems need to be explored carefully. These should aim at providing the flexibility to balanced US doctrinal requirements at Corps level and below with Alliance operational necessities at Army Group level and above.

Doctrine. Only initial measures have been taken to examine the significance of the doctrinal encroachment problem within NATO. In the continuing discussions of doctrine at various national, service, and field command levels, careful exploration of the area of influence concept and revised air allocation procedures should be undertaken in the context of European exercises and follow-on discussions. In particular, US Corps initial areas of influence should be developed based on both general guidelines and also specific terrain features, road nets and enemy capabilities, and the impact of these areas on other national Corps should be examined.

Due regard should also be paid to the development of intelligence for the non-US Corps. This may require compensatory efforts at Army Group or Allied Tactical Air Force (ATAF) level to assure a balanced interpretation of the battlefield in all Corps areas.

Finally, representative ground maneuver plans should be subjected to sensitivity analysis with regard to available air interdiction support. The right balance must be struck between providing adequate air interdiction to support the ground commander's scheme of maneuver to assure that air allocation achieves the most decisive effects theater-wide. As yet, it is too early to determine whether the SHAPE and AirLand Battle approaches can be harmonized to achieve this balance.

Materiel. The two approaches diverge somewhat with regard to their implications for the development and acquisition of materiel and both are somewhat ambivalent with respect to the extent to which they rely upon technological advancement and future procurement.

The SHAPE approach requires area surveillance, target acquisition, and attack systems of somewhat greater range. AirLand Battle, in addition to not requiring systems capable of the longer ranges, might entail greater attention to tactical C^3I, ground mobility, and logistic preparation of the battlefield to support more maneuver oriented warfare and greater air-ground synchronization.

Although both AirLand Battle and the SHAPE concept claim to be viable with current weapons and technologies, official and press discussions of them make constant references to the opportunities afforded by enhanced technologies. Most of such references are to area surveillance, target acquisition, and attack systems still in development and not expected to be available in significant numbers for another five to ten years. Moreover, it will be necessary for very complex developments in all functions—surveillance, information processing and dissemination, munitions, and weapon systems integration—to come on line successfully before their synergistic effect provides the anticipated leap in operational capability required for some of the advanced concepts of deep attack (especially those involving attacks against units as opposed to fixed facilities) to work to full effect. Given the procurement costs, technological uncertainties, time lines, and interservice issues involved, it is a very high risk force development process.

One of the greatest dangers is that we may become so enchanted with the future potential of the emerging technology that we (a) fail to continue procurement of proven systems and incremental upgrades necessary to carry us through in the interim, and (b) that we build current plans, strategy, and tactics as though the potential capability already exists.

The materiel and force development implications of AirLand Battle and the SHAPE concept are competitive but not necessarily incompatible. While no choice between these two doctrinal approaches seems otherwise required yet, it is imperative at this early stage in the development of deep attack systems to give consideration to the problem of doctrine and its relationship to force structure and materiel development. It is essential to confront directly the problems of attacking very deep with nonnuclear systems, lest we inadvertently set our sights on capabilities whose expenses and technical difficulties compound rather than ease the more immediate challenge of strengthening NATO's forward defense.

ENDNOTES

1. Phillip A. Karber, "To Lose An Arms Race: The Competition in Conventional Forces Deployed in Central Europe, 1965-1980" (Unpublished Paper, Washington, DC: 1981), pp. 65-67.

2. Robert S. Cooper, "The Coming Revolution in Conventional Weapons," *Astronautics and Aeronautics,* vol. 20, no. 10 (October 1982), pp. 74-75; and Benjamin F. Schemmer, "NATO's New Strategy: Defend Forward, But Strike Deep," *Armed Forces Journal International,* vol. 120, no. 3 (November 1982), p. 59.

3. US Army, FM 100-5, *Operations,* (August 20, 1982), p. 1-1.

4. *Ibid.,* pp. 7-1 thru 7-17.

5. *Ibid.,* p. 7-15.

6. *Ibid.,* p. 7-11.

7. Interview with Colonel Kenneth Keller, Special Assistant to the Chief of Staff, Supreme Headquarters, Allied Powers Europe, SHAPE, Belgium, February 9, 1983.

8. Julian Critchley, "Defending Europe by Attacking Attackers," *Daily Telegraph,* November 1, 1982, p. 18, quoted in *Foreign Broadcast Information Service Daily Report (Western Europe),* November 12, 1982, pp. Q8-9 (hereafter cited as *FBIS (Western Europe).*

9. Schemmer, "NATO's New Strategy: Defend Forward, But Strike Deep," p. 51.

10. General Meinhard Glanz, Interview in *Stern,* November 11, 1982, pp. 294-307.

11. Speech by General Bernard W. Rogers, Supreme Allied Commander Europe, Brussels, September 1982.

12. "A New Atlantic Strategy," *Le Monde,* October 5, 1982, p. 1, translated in *FBIS (Western Europe),* October 7, 1982, p. K2.

13. *Ibid.*

14. McGeorge Bundy, *et. al.,* "Nuclear Weapons and the Atlantic Alliance," *Foreign Affairs,* vol. 60, no. 4 (Spring 1982), pp. 753-768.

15. Cooper, "The Coming Revolution in Conventional Weapons," p. 74.

16. David S. Yost and Thomas C. Glad, "West German Party Politics and Theater Nuclear Modernization Since 1977," *Armed Forces and Society,"* vol. 8, no. 4 (Summer 1982), p. 528.

17. Walter Pincus, "Army Would Like Advance Authority to Use A-Weapons," *Washington Post,* July 21, 1982, p. 18.

18. Glanz, interview in *Stern,* pp. 294-307.

19. FM100-5.

20. Colonel Bud Adair, US Army, "AirLand Battle Briefing," Fort Monroe, Virg.: US Army Training and Doctrine Command, November 1982.

CHAPTER 5

NUCLEAR-CONVENTIONAL TRADEOFFS:
THE DEBATE IN EUROPE

by

Catherine McArdle Kelleher

In the early 1980s, the continuing Atlantic debate on common security and survival issues turned increasingly to the questions of conventional defense and deterrence.[1] The issues raised were many and contentious: Were there new conventional options or merely new methods and technology towards old ends? Were there new requirements for resource allocation and force structure development within and among the allied states? If so, how much would the new conventional options cost and what consequences would they have for deterrence?

The argument here is that the debate is compelling but of secondary importance without major economic, political, and social changes within the broader Atlantic context. Popular interest and the level of expert argument, particularly in Europe, is as intense as it has ever been in NATO's history. But Alliance development of independent, innovative conventional options first will have to surmount tests of economic feasibility and political acceptability, as well as military credibility. And the way forward, again especially in Europe, is neither clear nor without serious risk.

This chapter will focus on the nature of the European debate, and the problems of political logic and economic constraint it reveals. Attention will be given first to the causes of increased interest and commitment to conventional alternatives and then to the actual range of opinions among the key NATO partners. In conclusion, the factors critical for policy change will be highlighted and propositions developed for delimiting conventional options in the near future.

THE PRESENT CONTEXT:
NEW INTEREST, OLD ARGUMENTS

In essence, the search for greater conventional options is neither a new concern for NATO nor one which has gone unrecognized in NATO force structure or declaratory doctrine. Every Supreme Allied Commander Europe (SACEUR) since General Dwight D. Eisenhower has called for greater conventional strength and preparedness; most NATO Council meetings have emphasized the need for the West to do more in the face of Soviet conventional superiority and the stakes involved in any escalation to nuclear use. Moreover, the largest portions of all NATO military expenditures have gone to conventional forces. Indeed, the conventional share has absorbed a rising proportion of all NATO expenditures over the years since 1970 whether calculated in constant units or as a percentage of the national Gross National Products.[2]

What is new is the increased willingness, on the part of European elites and publics, to consider what constitutes an adequate, sufficient conventional option. Ten, even five years ago, conventional requirements were subjects barely touched upon in most European discussions. The standard dogma—if indeed the issue was ever raised—involved three interlinked propositions. First, the Warsaw Pact's conventional position in time of "hot peace" was invincibly superior. Second, the West was unable, if not unwilling, to mobilize more conventional assets except in war, when hopefully there would be sufficient warning. Third, there was a high probability that the conventional phase of any serious European conflict would be measured in days or hours given the importance of the stakes involved and the integration of nuclear weapons into standing ground and air forces. Therefore, nuclear use was perceived to be inevitable and escalation to all-out war, practically unavoidable.

Underlying these arguments was what has often been described as a logical puzzle. Nuclear war, particularly in Europe, was simply unthinkable; conventional war as witnessed from 1942 to 1944 was clearly unacceptable. The basis for defense planning, therefore, must be deterrence and the avoidance of any rhetorical or operational signal that conflict, either nuclear or conventional, could be tolerated or confined. The one operational doctrine which did receive emphasis was the requirement for the forward defense of West German territory, the *sine qua non* of Germany's continued membership in NATO.

Without question, there was significant dissonance between these belief structures and the sizable conventional efforts mounted at least by the major European states. Germany has had at least 400,000 under arms since 1965; even the smaller states maintained fairly constant manpower levels throughout the 1970s. To justify these manpower levels, European governments explained to generally quiescent publics and parliaments that such substantial European contributions were needed to insure the United States' extended deterrence guarantee. The terms of this guarantee had been laid down in the early 1950s; any change risked Congressional reconsideration of the NATO guarantee; US economic and political pressure in areas essential to European prosperity; and the unraveling of the basic postwar political balance in Europe, especially the encapsulation of the problems of German military power. Should specific military burdens pose undue economic and political strain, there were always interim political solutions through the bilateral and multilateral negotiations of NATO's Annual Review procedure. Often there were also offsetting side payments which would appeal to domestic pressure groups interested in technological advancement, employment opportunities, or long-term economic investment.

The Grounds for Present Interest. Every opinion indicator— from the types of stories covered in the daily press to nationwide opinion polls—suggests that the period of silence on issues of military policy is over. There is widespread dissemination of information on all military subjects and on nuclear war in particular. There is increased willingness to identify security issues as among the "most important" at present, and a new sensitivity on the part of voters to partisan political positions in this area. Nuclear war and nuclear strategy clearly command the greatest share of attention at all levels of knowledge and emotional involvement. But there is growing concern about all issues of security—the economic choices involved, the sharing of tasks and risks within NATO, the prospects for arms limitation in the future, to name only the most salient.[3] This public interest is not expected to diminish soon.

What has impelled this dramatic rise in attention and concern? Fear—of nuclear war, of new nuclear risks, and of imminent danger to all states—is clearly among the most significant motive causes. Estimates vary, but expressions of growing popular fear in Europe began in 1978-79 around the time of the widely-reported enhanced radiation or neutron bomb imbroglio. European concerns were fed by many factors—the irritation and frustration growing out of broad opposition to the Intermediate Nuclear Force (INF) deployments, the risks attendant to the situation in Poland, and the common European perception that President Reagan sought direct confrontation with the Soviet Union in all things, but particularly in the nuclear arms race. The tactical successes of the anti-INF push and of the more general antinuclear movements mobilized new groups to oppose the politics of the status quo. In at least the short run, they also resulted in a deepening of popular fears. They provided new factual bases for anxiety and an organizational basis for the coherent public articulation of previously private anxieties and questioning. The links across European borders led to significant coordination and common agenda-setting. By 1982, the emergence of the freeze movement in the United States and the pastoral letters of the American Catholic bishops added new legitimation and new political dimensions to demands for policy reconsideration and change.

A somewhat different framework for these fears has been the feeling of general pessimism, the perceptions that present societal problems are intractable or not subject to past remedies. Pessimism has been most significant in the high areas of unemployment throughout Europe, especially in "old" core industries (e.g., steel, coal, autos) and especially among school leavers and the under thirty generation, the tail end of the European baby boom. The available evidence is not conclusive. But there is much to suggest that general negativism about personal futures has led both to pessimism on other more serious problems (e.g., the fate of the nuclear world) as well to general distrust of governmental capability to solve any present issue. The conclusion of many economists that European growth rates will continue to lag throughout the next decade and that structural unemployment will be at least as serious as that in the United States, therefore, suggest a continuing negative trend.

A third, related cause is reflected in the major changes in leadership and policy which have taken place in England (Thatcherism), France (a conservativist Socialist government) and the Federal Republic (Kohl's election as a move toward the Right and away from detente). Each of these has mirrored and exacerbated Right-Left polarization—on security issues and domestic policy directions. The shifts in governing coalitions mean defense is again on the agenda for national debate. Discussion no longer threatens coalition harmony (as under Schmidt) or leads to ideological stalemates (as now under Labour or even perhaps to a future Social Democratic Party-Liberal government) which limits the ability to govern.

Moreover, these new governments have the benefit of the Alliance's usual learning curve.[4] Broadly described, this is the eighteen to twenty-four month period it usually takes the allies, and especially the European states, to absorb new information and to refashion American positions to their own domestic political requirements. The percolation effect, for example, of the concepts of AirLand Battle and AirLand Battle 2000 (now redesignated Army 21) is now fairly complete. The full dimensions of General Bernard W. Rogers' initiatives (Follow-On Force Attack) and the various expert proposals (American and European) on emerging technologies are visible and available for debate.[5] This does not ensure final decision or even agreement with positions developed by the United States. But it clearly lowers both the anxiety level of officials and challengers alike, and leads to a more expert discussion within governments and among allies.

The Spectrum of Present Views. To describe fully the range of European views, including those of governments and significant

Figure 5-1
SCHEMATIC OF SPECTRUM OF EUROPEAN VIEWS

1. ———— Conflict miniscule probability; Soviets self-deterred from nuclear use; nuclear and conventional to be reduced

2. ———— Conventionalization the only hope, sufficient deterrent

3. ———— More conventionalization needed, even for deterrence

4. ———— Better nuclear-conventional balance needed but both necessary for deterrence and defense (e.g. second echelon)

5. ———— More conventional options only where nuclears manifestly incredible (BNW's) or risky (ADM's)

6. ———— Conventional enhancement to meet popular concerns; primary concerns deterrence through threat of nuclear escalation

7. ———— Nuclear modernization and expansion essential: middle "gap" and need for greater autonomy/ European role

opposition groups, is beyond our scope here. Our attempt rather is to develop a framework within which to describe these views systematically and to allow for broader comparison. The instrument will be the schematic depicted in Figure 5-1, a synthetic classification of positions in terms of the importance accorded the pursuit of greater conventional options, and the actual or implied beliefs about nuclear-conventional tradeoffs.

Figure 5-2 endeavors to show the range of "centrist" opinion in some major NATO states as well as the views of prototypical groups (Campaign for Nuclear Disarmament (CND), the "Greens") and the official NATO position (including the Supreme Headquarters Allied Powers Europe (SHAPE) variant). Even given the necessary simplicity of this depiction, two conclusions are clear. First, there is far greater interest in a new range of conventional-nuclear tradeoffs, in terms of both direct replacement in functional roles, and redirection of effort in the allocation of funds, men, and

Figure 5-2
INTRA- AND INTERNATIONAL DIFFERENCES: AN APPROXIMATION

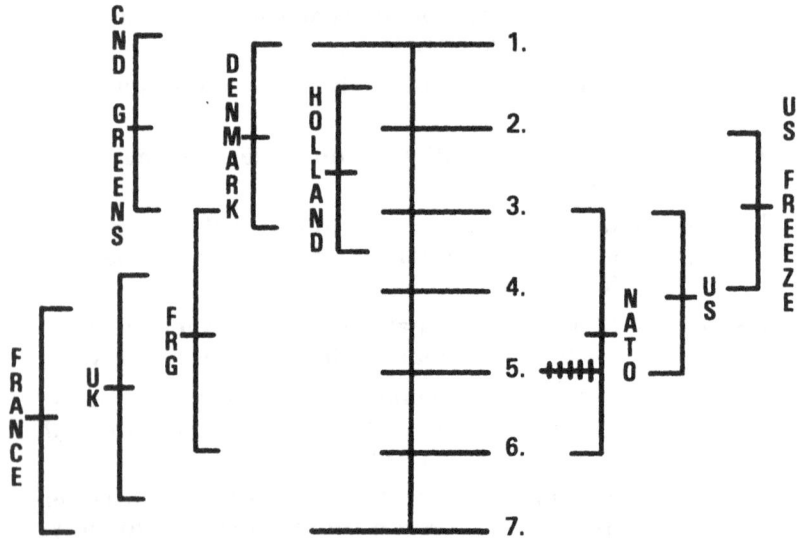

—— MIDPOINT OF OPINION RANGE

┼┼┼┼┼ NATO SHAPE

materiel. Second, the dominant set of official opinions still views the conventional option as an "add-on," a set of capabilities which will be valuable additions if they can be procured at acceptable cost. They are not replacements for nuclear weapons, in deterrence or defense—generally or in the majority of specific cases beyond those of battlefield nuclear weapons (BNW) or atomic demolition munitions (ADMs).

Also, Figure 5-2 shows that significant differences exist between the smaller states and the major land powers, and between the European nuclear haves and have-nots. The span of national divergences is striking. It is not so great as to destroy the general doctrinal parameters set down in NATO's MC 14/3, the basic 1967 statement of flexible response. However, this range of difference does allow for the development of consensus under a US leadership, particularly if it is willing to compromise at the margins. But the political tasks involved would seem formidable ones, certainly for the Reagan Administration in its present posture but also for any conceivable successor over the next decade.

FACTORS CRITICAL FOR POLICY CHANGE

What in the foreseeable future are the conditions which will stimulate or constrain efforts to increase a conventional emphasis in both strategy and force structure? Three key factors will affect a range of national decisions in Europe. All would militate against greater conventionalization (except perhaps at the rhetorical level), unless accompanied by major (and presently unexpected) political and economic changes.

The Complex Character of Public Views. First, the nature and intensity of public pressure toward conventionalization are significant. Evidence of widespread antinuclear sentiment, of unease with the present level of dependence on nuclear weapons, abounds. Popular opinion has been mobilized; there is every reason to expect that the issue will remain on the popular political agenda, even if direct mobilization slackens off. What is not clear is the degree to which antinuclear sentiment will translate directly into greater enthusiasm for conventional preparedness and new levels of conventional expenditure. No such transformation has occurred in past periods of nuclear anxiety (1957-58 in Germany or the CND in Britain until 1961). A clear signal of leadership, able to put forward credible options in ways which are both persuasive and comprehensible, will be needed to capitalize on the antinuclear sentiment. No government has yet undertaken this task, although clearly it is at the top of the agenda in the Defense Resource Review in Germany and the subject of continuous discussion in Denmark and Holland. However, the leading role will in all probability fall to opposition groups as in the 1950s. Not surprisingly, the SPD with Egon Bahr's Strategy Commission has taken the initiative and produced a number of explications of conventional strategy. The challenges of finding a range of acceptable solutions and then insuring party support, let alone alliance credibility, have yet to be met.

The profile of public attitudes also includes several contradictory tendencies, both within national populations and across the various European states. Antinuclear stands in principle attract majorities everywhere but France. Opinion polls show that Europeans foresee nuclear use only in situations of dire national peril, if indeed at all. Yet at least pluralities favor retention of deterrent strategies (even those based on nuclear weapons) especially if the choice is development of warfighting strategies. The paradoxes among the British and French response patterns are even more pointed. More than sixty percent of each national sample favors retention of the national nuclear deterrent. Almost equal percentages refuse to endorse the use of American weapons from their soil; there is strong majority support in England for their immediate withdrawal.

The specific aims of public pressures appear even muddier when the cost factor is introduced. The general expert perception is that the conventional option is one which will impose major new economic burdens at just the wrong time. The barriers appear formidable. The West European states have witnessed a marked decline in real economic growth; their lack of competitive edge in certain high technology sectors is seen with considerable foreboding. Energy supplies are no longer as grave a problem for consumers (Germany, France) as predicted, but neither are they proving as much of a national bonanza for suppliers or refiners (Britain, Norway, Holland). No European state has met the NATO three percent growth targets in their recent defense budgets; no one has hope of significant near-term improvement.

Moreover, a program of conventional improvement now would come just at the close of a relatively successful package of general modernization measures under the NATO Long Term Defense Program (LTDP). While wish lists abound, few European military leaders or their political counterparts are willing to wage the political and economic battles that would be required to justify new priorities and new investments in materiel, facilities, and general sustainability measures. The political costs almost certainly would be high in political environments which are already sensitive to the needs generated by high youth unemployment and issues of structural obsolescence in industrial plant and skill maintenance.

Demographic Factors. The next decade will indeed see a number of evolutionary changes which are almost certain to push the costs still higher. The first are the widespread shifts in European demographic profiles reflecting the introduction of reliable birth control in the 1960s, as well as the effects of better medical care and economic security among the aged. Fewer and fewer men will enter the conscription cohort from now until the end of the century. If exemptions for conscientious objectors and for those from special family circumstances or the otherwise "socially disadvantaged" continue, there will have to be more rigorous call-ups and extensions in terms of service to maintain present manpower levels.[6]

In the competition for scarce resources, men, and money, the military will be at a disadvantage given the generally lower social standing it enjoys in all the postwar European societies.[7] There has been a measurable increase since the nadir of popular standing (the late 1960s) when the effects of Vietnam disaffection and the rise of a popular counterculture were most pronounced. There is also some appreciable rise in interest in a secure career in troubled economic times, especially if present trends in youth unemployment continue unabated. But the most effective means toward conventionalization require even larger numbers of trained personnel than at present and it is precisely these skills which will be in most demand throughout the European societies.

Each of the European military establishments have looked at ways to avoid these extra costs by using untapped personnel pools. In many respects, the analysis is farthest advanced in Germany which will suffer the earliest, most severe effects of demographic change. German studies to date have considered several major target groups: especially (1) women; (2) immigrants and the German children of immigrants (e.g., Turkish immigrants of the 1950s and other guest worker stay-behinds); and, (3) socially-disadvantaged groups needing educational and medical help to attain military standards. Conscripting sufficient numbers from each group would bring with it high initial start-up costs and major requirements for social and educational changes. So far, at least, these costs have seemed too high for the numbers gained. Unless unemployment continues, however, the first critical shortages will come in 1986. There must be some modification in conscript practices if total Bundeswehr numbers are not to fall below the critical "threshold" of 450,000.

The Superpower Connection. The third factor is European political calculations about their long-term relations with the superpowers. Clearly, any significant move toward conventionalization will raise questions about extended deterrence and present Atlantic foreign policy convergence. Already, there is widespread concern that conventionalization will speed US decoupling and what may be, but is not yet necessarily, an ultimate withdrawal from Europe. If this were to occur, not only would there be inadequate substitution of a conventional deterrent for a nuclear deterrent; there would be even more certainty within Europe that, under parity, the United States would be unwilling to trade Hamburg, Dusseldorf, or even Amsterdam for the sanctuary of its continent. In political terms, protection based even partially on the British and French nuclear systems brings high costs for Europeans, especially Germans—and no more certainty, among the mass or elites. All of the political burden implied in Adenauer's nightmare of European footsoldiers in bondage to far-off American nuclear knights would come true. And there would be a *de facto* shift in the economic costs of defense, as well as in alliance responsibility.

Vis-a-vis the Soviet Union there would seem a greater range of contingencies. In either real terms or for propaganda purposes, the Soviets might declare conventionalization to have raised the specter of imminent war or direct preparation for warfighting in the near term. Particularly given what of necessity would be the prominent German role, there would be useful propaganda images for the Soviets of "Germans again on the move"—images ceremonially buried under Ostpolitik but still viable politically in Eastern Europe, in the Soviet Union, and even in some West European circles. It is also not clear how conventionalization would affect what few chances remain for significant arms control in Europe. Would it lead to greater stability given the visible nature of the buildup? Provide new incentives for mutual force reductions? Allow for an even larger number of confidence-building measures? The evidence is far from clear; suspicions based on past history and the present frozen state of relations would suggest less than optimistic outcomes.

A EUROPEAN PROPOSITIONAL GUIDE
TO CONVENTIONAL OPTIONS

At this point, therefore, a general conclusion must emphasize that few Europeans await the near-term implementation of greater NATO conventionalization. Some oppose it directly and vigorously, as in recent SHAPE bilateral discussions regarding both AirLand Battle concepts and Follow-On Force Attack. Most expect the ideas, while attractive, to stagnate given the political and economic barriers described above.

Those who hold a "wait-and-see" position might be said to be guided in their considerations by the following six broad propositions. All are worth considering; no one alone will be decisive. Proponents and opponents of conventionalization would do well to heed them.

• Western defense against the first Soviet-East European echelon is at least as important as new deep attack ideas about the second and follow-on echelons.

• Adoption of an offensive or preemptive strategy by NATO in peacetime is politically unacceptable to most West European governments and peoples.

• NATO's commitment to a doctrine of forward defense is an inevitable compromise between a mobile defense made impracticable by space and density considerations, a defense in depth which is politically and militarily impossible, and the political realities imposed by West Germany's NATO membership.

• Emerging technologies, especially those of interest among conventionalization advocates, are promising solutions which are not yet certain and may not be acceptable.

• Plans to develop new reserve resources or to restructure some standing forces along cadre lines may be desirable in terms of cost, but not in political or military effect at home and perhaps to ally and opponent as well.

• Specialization or a new division of roles, missions, and structures among Atlantic partners may be the wave of the future but how will NATO begin?

ENDNOTES

1. The research on which this chapter is based has been generously funded by a NATO fellowship and by a grant by the International Security Affairs program of the Ford Foundation to the author, William K. Domke, and Richard C. Eichenberg. The first major publications include the co-authors' article "The Illusion of Choice: Defense and Welfare in Advanced Industrial Democracies," *American Political Science Review,* vol. 77, no. 1 (March 1983), pp. 19-35; and the collected essays on national elite opinions and foreign policy profiles in Wolf-Dieter Eberwein and Catherine McArdle Kelleher, eds., *Sicherheit—Zum Welchem Preis* (Munich: Olzog, 1983).

2. See for this data Caspar Weinberger, *Report on Allied Contributions to the Common Defense* (Washington: US Government Printing Office, 1981, 1982, 1983).

3. Compare the treatment of national opinion profiles in David Capitanchik and Richard Eichenberg, *Defense and Public Opinion,* Chatham House Papers in Foreign Policy (London: Routledge and Kegan Paul, 1983) and in Steven Szabo, ed., *The Successor Generation* (London: Butterworths, 1983).

4. See an earlier treatment of this subject in Catherine McArdle Kelleher, *Germany and the Politics of Nuclear Weapons* (New York: Columbia University Press, 1975).

5. Beyond the ideas in this volume, references include General Bernard W. Rogers, "The Atlantic Alliance: Prescriptions for a Difficult Decade," *Foreign Affairs,* vol. 60, no. 5 (Summer 1982), pp. 1145-46; and *Strengthening Conventional Deterrence,* Report of the European Security Study Group (New York: St. Martin's, 1983).

6. See, for example, earlier projections made in the remarkable series of German *White Books on Defense,* including the latest one published in 1983 (Bonn: Ministry of Defense, 1975/76, 1979, 1983), as well as the groundbreaking report of the Force Structure Commission of the early 1970s commissioned by then Defense Minister Helmut Schmidt (Bonn: Ministry of Defense, 1972).

7. Compare here Francis Pym "Defense in Democracies: The Public Dimension," *International Security,* vol. 7, no. 1 (Summer 1982), pp. 40-44.

CHAPTER 6

THE ANATOMY OF THE SOVIET EMPIRE: VULNERABILITIES AND STRENGTHS

by

Vernon V. Aspaturian

Although the deployment of American Pershing IIs in West Germany and Ground Launched Cruise Missiles (GLCMs) in Britain, Italy, and West Germany has determined, for the short run at least, that Western Europe will continue to rely upon American nuclear power to deter a Soviet military move westwards and thus shield Western Europe from intimidation and political submission, the problem of defending Europe militarily if deterrence fails remains uncertain. West Europeans have made it quite clear that they do not wish to be defended against Soviet military assault with nuclear weapons, whose function has been perceived to deter war, not win it. With the achievement of strategic military parity by the Soviet Union and possibly regional nuclear superiority in Europe with the deployment of the SS-20s, many Europeans were becoming convinced that the United States would not put its own cities and populations at risk by responding to a Soviet attack in Europe with nuclear weapons launched from the United States.

The NATO two track decision adopted in 1979 was designed first to deter further Soviet deployment of intermediate range missiles (SS-20s) and indeed to persuade Moscow to dismantle them through negotiation. If negotiations failed, NATO would proceed to counter the Soviet deployment with a force of American intermediate range missiles, made up of 108 Pershing IIs (based in West Germany) and 464 cruise missiles based in West Germany, Italy, Britain, Netherlands, and Belgium.

Although a two track decision was designed to reassure West Europeans that a European based American INF capability would be more effective in deterring the Soviet Union, many West Europeans increasingly felt that it increased the possibility of nuclear war if deterrence failed, since the United States might use its European nuclear force to localize a nuclear conflict in Europe. Imperceptibly, American strategy shifted from the concept of nuclear deterrence to nuclear defense, which came to a head during the early months of the Reagan Administration when high ranking American spokesmen, including the President himself, publicly ruminated about the possibility of limiting a nuclear conflict to Europe in the event deterrence failed. Antinuclear sentiment and movements accelerated in Europe, especially in West Germany and Britain, where they gained the support of the Social Democrats and the Labour Party. The Labour Party actually adopted a platform of denuclearizing Britain completely, and the German Social Democratic Party reversed itself by voting overwhelmingly in the German Parliament against the deployment of American missiles in West Germany.

THE EUROPEAN NUCLEAR DILEMMA

Michael Howard has persuasively argued that the American umbrella was designed primarily to provide Western Europe with reassurance that its power would be sufficient to *deter* a Soviet military attack, not to defend it with nuclear weapons.[1] The reassurance which American nuclear weapons provided was the guarantee that their existence would prevent nuclear war, not unleash it. As long as the United States was perceived as possessing overwhelming nuclear superiority, the magic of deterrence in Europe continued to function; but once the Soviet Union achieved global parity and European superiority in nuclear weapons, the American nuclear umbrella was perceived as increasingly promising not deterrence, but a nuclear defeat for the Soviet Union on European soil in the event deterrence failed, a prospect that hardly served to reassure large sectors of West European publics and indeed a substantial share of American sentiment as well.

If negotiations with the Soviet Union fail to resolve the Euromissile issue and American deployment of Pershing IIs and cruise missiles continues, the credibility of even an American Euromissile force may erode in the face of growing protests, as the issue increasingly becomes one of avoiding nuclear war altogether. This has stimulated a search for nonnuclear options. These options in turn all presuppose that, if the West refrains from using nuclear weapons, the Soviets will do likewise. But as long as nuclear weapons exist, their use is possible, and no amount of unilateral self-restraint can guarantee their nonuse. Thus a nonnuclear deterrent must be sufficient simultaneously to threaten enough punishment to make a Soviet military move unprofitable, or if such a deterrent fails, to repulse a Soviet military assault without provoking the Soviet side into using nuclear weapons to prevent defeat. Otherwise, the entire conception of a nonnuclear option collapses. This is indeed a tall order.

Samuel Huntington, in a provocative essay, "The Renewal of Strategy," offers a nonnuclear strategy which once again seeks to recouple reassurance with both deterrence and defense.[2] Huntington calls for a nonnuclear retaliatory force that would be sufficient to deter a conventional military attack and to seize Soviet assets in Eastern Europe without resort to nuclear weapons if deterrence fails. For such an alternative to work, a number of assumptions must be operative, among the most important being that West Europeans, especially Germans, would perceive a conventional war between NATO and the Warsaw Pact forces as less destructive than a nuclear war. For the Germans, at least, this assumption is somewhat dubious.

Although one may quarrel with the specific strategy which Huntington has proposed—an offensive conventional NATO force which would immediately retaliate by driving forcefully and instantaneously into East Germany and Czechoslovakia to seize something of value to Moscow—he has nevertheless generated a broader conceptual spectrum of alternative strategies to American nuclear retaliation. These alternative strategies need not be restricted to nonnuclear options, and allow for the creation of purely European alternatives, a number of joint American/European alternatives restricted to Europe, an American global alternative, and an American global alternative involving European participation. Huntington also explores varieties of nonmilitary deterrence strategies involving either the United States alone or in conjunction with Allied efforts. Furthermore, Huntington conceptualizes a Soviet empire divided into two zones, an inner, vital zone and an outer, peripheral zone, and examines a number of ways to threaten the assets of the Soviet Union through "weaning" or "weakening" measures, tailored for individual components of the Soviet empire.

In another essay, Huntington distills his conceptual ruminations into four possible NATO strategic alternatives: (1) continuation of existing strategy and forces; (2) creation of a massive conventional defensive force; (3) creation of a German nuclear deterrent; (4) adoption of a strategy of conventional retaliation. Huntington offers the view that "there is unlikely to be an immediate consensus forming in support of either of the latter two possibilities," because both are likely to be viewed as "unsettling and dangerous."[3] One may add that the alternative he prefers is indeed "unsettling and dangerous," not because it is a new idea as Huntington claims, but because it is simultaneously threatening and ineffective. But more about this below.

Huntington appears to favor strategies that involve both the United States and Western Europe, that are nonnuclear in character, and are decoupled from Soviet-American global rivalries elsewhere. These are, of course, concerns central to European perspectives: American involvement in Europe, European noninvolvement in Soviet-American global competition, and exclusion of nuclear war in Europe. He hints at a purely European alternative, i.e., a German nuclear deterrent capability as a second choice, but ignores the possibility of other European nuclear alternatives, to wit: (1) a modernized and expanded British and French nuclear capability completely decoupled from the American nuclear force sufficient to establish an independent nuclear parity with the USSR in Europe; (2) an independent NATO nuclear deterrent capability involving West Germany and Italy and possibly the smaller West European powers. Both could serve to decouple Europe from the American nuclear force, insulate Europe from Soviet-American rivalry elsewhere, and allow Europe to make the ultimate decisions concerning its fate. These alternatives would give up the assurance of American involvement, but at the same time they would be more effective and credible than a nonnuclear retaliatory force involving the United States. Thus, whereas an American Euromissile retaliatory capability is likely to be perceived as excessive and hence dysfunctional, a purely European conventional retaliatory force undesirable and inadequate, and a joint US-European conventional retaliatory force desirable but insufficient, a British-French or European NATO nuclear parity with the Soviet Union, with American support only in reserve, would appear to be the most effective militarily but among the most difficult to develop politically.

As noted earlier, a successful alternative must deter, reassure, and defend, but this calculus is both infinitely subtle and multidimensional in nature. The impact upon all three actors involved in the equation (the United States, Western Europe, and the Soviet Union) must be examined very carefully. Eastern Europe is not calcuated into the equation because it is not an independent actor, but remains essentially an instrument of Soviet policy and an object of Western policy. A successful policy, whether exclusively European or joint American-European, must deliver the following:
- deter the Soviet Union,
- reassure Western Europe,
- reassure the American public, and
- reassure the Soviet leaders.

Reassurance here means exclusion of nuclear conflict.

Irrespective of the strategy employed, it is important to note that the Soviet Union is less likely to resort to war to expand its assets than to defend them, and less likely to run risks protecting its assets than in securing its survival.

Hence, the Soviet Union under existing conditions is not likely to employ force to expand its assets in Europe, will be quick to use conventional military means to protect its European assets, and is likely to resort to nuclear weapons only when the assets threatened are perceived to be of sufficient magnitude to endanger Soviet survival. Correspondingly, past Soviet behavior strongly suggests that it will relinquish assets that are marginal in value or peripheral to its central concerns rather than incur the risks involved in using force to protect them (Egypt, Iraq, Grenada, etc.). It will also surrender assets of this character in search of wider political and diplomatic gains (giving up bases in Finland and assets in China; signing the Austrian State Treaty). Furthermore, it will also surrender assets in order to acquire or protect more important assets (Somalia for Ethiopia), lessen a threat to its survival or to maximize its chances for survival (concessions to Hitler in the months before the German attack).[4]

Although the Soviet Union's most valuable assets are in Europe, the Soviet Union is no longer a simple, extended territorial empire, but is now a global power with allied and client states in various parts of the world distant from the Soviet Union. Although Europeans have become almost obsessively concerned about being involved in Soviet-American global rivalries, the global character of the Soviet imperial system inevitably involves it in activities that can seriously affect the interests of Western Europe. To date, the West Europeans have been content to allow the United States to protect their interests in other parts of the world against Soviet encroachment while resisting involvement. A purely American defense of West European interests around the globe is increasingly becoming as difficult as a purely American nuclear deterrent in Europe.

Just as Soviet strengths have proliferated globally, so have its vulnerabilities. Accordingly, the defense of Western Europe may conceivably be more effectively enhanced by threatening assets outside Europe, which may be both valuable and vulnerable in contrast to Europe itself, where Soviet assets are indeed valuable but also considerably less vulnerable, and hence more dangerous to threaten.

NATURE OF THE SOVIET EMPIRE:
A SYSTEM IN SEARCH OF CONCEPTUALIZATION

In order to properly inventory and assess the comparative value, vulnerability, exposure, and defensibility of Soviet assets throughout its empire, it is first necessary to conceptualize the Soviet imperial system, comparing it with past empires, defining its components and elements, and relating them to one another and to the outside world. The problem is rendered difficult because the Soviet empire is still in a state of flux, expanding and contracting simultaneously, and still in search of its own definitive equilibrium. The Soviet empire has been growing both continuously as well as sporadically and amorphously. Needless to say, since Soviet writers would deny the existence of a Soviet empire and hence the validity of any conceptualization, we nevertheless discover provisional attempts to define a system of interrelationships between the Soviet Union and various nations, states, and domestic movements and constituencies within various countries that add up to a *de facto,* but officially unacknowledged, empire. Soviet concepts like "socialist community," "socialist commonwealth," "world socialist system," "socialist-oriented regimes," "international relations of a new type," together with a network of bilateral alliances, multilateral military, political, ideological, cultural, technical, and scientific organizations and associations with restricted membership, suggest a complicated imperial system made up of components interacting in varying degrees, levels, and intensities of involvement with the Soviet Union, the hub of the system, and with each other. And within the Soviet Union we find an inner hub, the Russian nation, which provides the national content of the system and is responsible for its energy and dynamism.

In conceptualizing the Soviet empire, one must inevitably begin with the term "empire" itself, to examine whether this label, with its enormous antecedent polemical and analytical preconceptions is still sufficiently explanatory in its cognitive capabilities to be useful in describing the Soviet global system. The term empire and its derivative adjective, "imperial," have been used to describe a wide variety of extended state systems over time, space, cultures, civilizations, and sociopolitical orders. Of course, in most past instances, the problem of conceptualization was irrelevant because most empires from Ancient Egypt to Bokassa's putative and evanescent Central African Empire (the most recent) were self-designated as "empires."

During the nineteenth century, the era of "cosmic conceptualization," the simple term "empire" with its relatively uncomplicated meaning was detached from its historical etymological origins. It became converted into a complicated and variegated set of theoretical formulations called "imperialism," which almost immediately assumed polemical and ideological, as well as analytical, incarnations, with attributes of both derision and repugnance.

In the twentieth century, the term imperialism became almost inextricably intertwined with the concept of "colonialism," whose relationship to its entymological progenitor, "colony," was similar to that of imperialism and empire. After World War I, traditional noncolonial empires of Europe disappeared from the scene and the remaining empires were almost exclusively colonial in character. Dependency was the thread of continuity in these semantic permutations. As a result, both colonialism and imperialism were increasingly described as systems of dependency relationships, whether the nature of the dependency was legal, political, economic, military, cultural or ideological, irrespective of form.

It is ironic that the two self-anointed anti-imperialist states in the world today, the United States and the Soviet Union, are characterized by many as the only imperial systems or empires left in the contemporary world. To be sure, the term empire with respect to the United States is more epithetical than analytical, but with respect to the Soviet Union, a strong case can be made that both "empire" and "imperialism" retain their empirical accuracy and analytical validity. Whether the terms "colonial" or "colonialism" are also applicable or appropriate raises more difficult questions, which will be elaborated upon below.

What is an empire? What is imperialism? There is no intention here to examine the various theories of imperialism, of which there are many, and possibly still others being devised. Rather, we will employ a simple working definition of imperialism and then empirically describe and analyze the Soviet system within its framework. Imperialism here is defined as extending, perpetuating or preserving the domination (whether through political, military, economic, ideological or cultural means) of one nation or state over other nations, peoples, states or territories. This definition is broad enough to embrace ancient empires, colonial empires, and various forms of structural and functional imperialisms and state-systems with hierarchical or hegemonial dependency relationships. Empire and imperialism are untidy concepts like many in political science, and any definition of empire and imperialism is likely to be conceptually defective. They are terms with an historical legacy and at any given time, even when applied to new phenomena, necessarily carry some residual baggage of this legacy which contributes to the methodological defectiveness. Furthermore, as noted earlier, empire and imperialism (as other concepts in political science) are often value-laden and these attributes are not always easily discarded. In our own day empire and imperialism are pejorative terms, whereas in the past they were laudatory. In the past, state entities proudly flouted the banner of empire.

Another important distinction is also in order at this point. Imperialism is not necessarily colonialism, but colonialism is a specific form of imperialism. Every empire and imperialism has a national content, which identifies the nation from which the ruling element is recruited, or whose ethos, values, norms and language permeate the empire in varying degrees of extent and intensity. Over time, the national content may be diluted and thus the national identification of empire becomes fragmentary or residual, as various nations and peoples within the empire are co-opted into the ruling elements.

The chief distinguishing characteristic of colonial imperialism is that it is a form of domination in which a sharp distinction exists between the ruling nation and subject nations in terms of race, culture, or religion, which is more or less permanently maintained. An empire or imperialism may be exclusively noncolonial in character, exclusively colonial, or a mixture of the two. The latter type of imperialism of which the Russian empire of the Tsars was a conspicuous example, is the most difficult to define with precision. The Hapsburg empire both before and after the *Ausgleich,* on the other hand, is an example of a noncolonial empire.

The simplest and most uncomplicated form of empire-building is territorial conquest of adjacent territories by a state or nation and their formal direct incorporation into an expanded state. But empires can also be more complicated and subtle in their structure, involving varying degrees and forms of dependency, autonomy, and clientage, with asymmetrical relationships of obligation and benefits between the center and the periphery. The most successful empires in the past, beginning with the Persian and Roman, through the Byzantine, Holy Roman, and Islamic empires, and ending with the British, were extremely flexible, improvisational and adaptable in their structure. The Soviet empire, more so than its predecessor, the Tsarist empire, continues this pattern in its own idiosyncratic manner.

The imperial relationship is terminated, at least nominally, when individual components of the empire and their territories are juridically separated into sovereign states, or when the individual nations in a fully-expressed choice decide to remain within a larger multinational or multistate entity (federation or confederation) in which citizenship is equal, universal, and uniform.

THE STRUCTURE OF THE SOVIET EMPIRE

It is at once obvious, given the historical, ideological, and polemical baggage which the terms "imperialism" and "colonialism" bear, that considerable controversy can be aroused in applying the terms to the Soviet scene. Objections may come from different directions. Soviet writers and Soviet apologists, of course, reject the concepts as ideologically incompatible with the concept of a socialist society, yet dissident Marxist-Leninists in Beijing and Belgrade have developed elaborate theories explaining the nature of Soviet imperialism and colonialism. Some simply equate the Soviet Union with its Tsarist predecessor in new garb and see Soviet policies and expansionism as modern extensions of Russian imperialism. Others view explanations of this nature as historical reductionism, and either deny the applicability of the concepts to the Soviet scene, or develop value-free or ideologically neutral concepts and theories of dependency and hierarchical relationships.

The source of this controversy stems from the complicated origins of the Soviet state and its evolution which provide ample empirical, but ambiguous, evidence to support not only different but diametrically opposed conceptualizations of the Soviet system. Some of the reasons for these differing appraisals are the following:

• The territorial congruence of the Soviet state and the Russian empire, which many view as a sufficiently *prima facie* case for identity of the two.

• The formal juridical termination of the imperial and colonial system through the transformation of the Russian empire into a multinational federation of juridically equal nations, whose citizens enjoy universal, equal and uniform citizenship. Preserving the "right to secede" in the Constitution was designed to emphasize the free association of the Union's component nations and sufficient to nullify any residual imperial or colonial coloration. These juridical transformations are accepted by some as sufficient to terminate the imperial and colonial relationship which existed prior to 1917.

- The transformations described above apply only to fifteen nations within the Union and are arbitrarily denied to other nations (whose populations may be larger), which are organized into four different and unequal levels of juridical-hierarchical national components. The fact of uniform citizenship alone is insufficient to nullify these unequal relationships. For example, although inhabitants of Puerto Rico and Guam enjoy uniform US citizenship, the United Nations still considers them "dependencies" of the United States.

- The continued existence of informal hierarchical distinctions among nations carried over from the Tsarist period. The Russians are quasi-legally established as *primus inter pares* among the nations, and given a central role in the Union as the "Elder Brother" nation and the nucleus of the Soviet state, as expressed in the national anthem. The fundamental national cleavage is between Christian and non-Christian (primarily Moslem) nations, and further gradations within each major category (explained below).

- The *de facto* domination of the Soviet state and Communist Party by the Russian nation, from which are recruited the overwhelming proportion of the leaders and decisionmakers, whose language and culture have been designated as "internationality" in character, and constitute the *de facto* official language and culture of the Soviet state. Historical attitudes towards the Russian nation by the other nations continue to provide a major reference point in defining the role and status of the non-Russian nations.

- The territorial expansion of the Soviet state as a result of the Nazi-Soviet Pact of 1939 and World War II, when entire nations and large tracts of territory were forcibly occupied, annexed, and incorporated into the Soviet state, at a time when other extended territorial states (whether allied or Axis) were in a process of territorial contraction.

- The historical origins of the Communist states of Eastern Europe, Mongolia, and North Korea, whose systems and, in some cases, statehood (East Germany and North Korea) were forcibly imposed by the Soviet Union which unilaterally exercises the right to intervene and preserve the Communist character of their systems and relationships with the outside world.

• The transitional and continuing expansionist and innovationist character of the relationship between the Soviet Union and other Communist states and with Third World countries, some of which are in varying degrees of integrated client and dependency relationships with the Soviet Union. What will now be called the Soviet empire is currently in the process of transforming itself from a territorially contiguous system into a global system with overseas components, and the status of the various components of the empire remain to be defined. The configuration of the Soviet empire is by no means terminal and it still seeks its definitive inner equilibrium.

The complexity of the situation is further exemplified by the inconsistent manner in which the Soviet empire is defined by those who employ the term. For some, the term applies only to the Soviet Union itself, and refers primarily to the relationship between the Russians and the non-Russians. Others use the term to include Eastern Europe; still others embrace other pro-Soviet Communist states; and finally, to an increasing degree, the term is also applied to include the relationship between the Soviet Union and various Third World countries as well. Even at this extended level, the situation remains untidy, because there still remains the relationship between the Soviet Union and nonruling Communist parties (all viewed as the nuclei of potential Communist states) and with dissident and even hostile Communist states.

Thus, in examining the Soviet empire in its current state, it is immediately apparent that it is made up of different components, with different historical associations and origins, and varying degrees of integration with the Russian nation which constitutes the national nucleus of the system. Three distinct components or rings of the empire can be delineated, with the Russian nation at the center. The first and most integrated component consists of the non-Russian nationalities within the USSR, which can be designated as the *inner empire.* The second component consists of other Communist states, which will be designated as the *outer empire,* the third component consists of Third World countries and will be designated as the *extended empire.*

The Soviet empire is still very fluid, although certain components and relationships are rigid and tight, while others are flexible and loose. Each component can lose or acquire members independently or at the expense of the others. The nature of the movement from one part of the empire to the other can also be varied in character. Thus, members of the outer empire were at one time viewed as potential members of the inner empire. Some former parts of the outer empire (Yugoslavia, Albania, China), have defected, although Yugoslavia's relationship with the Soviet Union is akin to being a member of the extended empire. Afghanistan provides a vivid illustration of the dynamic fluidity of the Soviet imperial system. Given the Soviet military occupation of Afghanistan, the country is being treated as if it were a part of the Soviet outer empire, whereas before the invasion it was a part of the extended empire. Its future status remains indeterminate: given the flexible character of the Soviet empire, it can wind up in any of the three components.

The entire Soviet imperial complex resembles a Russian Matroshka doll, with the Russian nation concealed and protected as the innermost doll by the inner empire of nations, which in turn is insulated and protected by the outer empire of Communist states, which in turn is cushioned by a band of socialist-oriented regimes and other clients in the Third World.

The Inner Soviet Empire. The inner Soviet empire, except for the loss of Central Poland and Finland and the gain of Carpatho-Rus and Eastern Galicia, is virtually territorially congruent with the Tsarist empire. For this reason, the Soviet Union is often described as the last of the nineteenth century colonial empires which has somehow managed to preserve itself intact while other colonial empires have been dismantled. The strong residual colonial character of the Soviet empire becomes evident when it is pointed out that the quasi-colonial portions of the Soviet state were for the most part forcibly incorporated into the Russian empire by conquest after 1860 during the heyday of European colonial expansion, of which Russian expansion was a major stream. The fact that Russian colonial expansion was at the expense of contiguous territories and included the simultaneous annexation of uninhabited or sparsely inhabited territories, independent states, areas claimed by the Chinese empire, and territories which were in effect *terra nullius* renders classification even more difficult.

These areas had been a part of the Russian empire for little more than fifty years at the time of the Bolshevik Revolution and thus were not yet fully integral parts of the Russian empire and in some cases retained their quasi-tributary status as protectorates of Russia. Initially, the Bolsheviks denounced conquests of these areas as manifestations of rapacious Tsarist imperialism and the annexation treaties with China were also denounced as unequal and illegal. However, after a brief flirtation with the idea of allowing the Russian empire to fragment in accordance with the principle of self-determination, Lenin backtracked and devised first the Russian Socialist Federated Soviet Republic (RSFSR) and then the expanded USSR as vehicles for preserving the territorial integrity of the Russian empire. Subsequently, under Stalin, Soviet historians reinterpreted the Russian conquests and retroactively declared that the territorial expansion and forcible incorporation of alien peoples into the Russian empire was an "objectively progressive process," because the Russian empire had been transformed into the Soviet multinational state. Moscow cleverly was able to convey the impression that these territories were in fact an organic and integral part of the Russian empire which the Soviet state was the juridical successor. As a consequence the Soviet Union has been spared the wrath of anticolonial ideologists and movements.

The inner empire retains strong residues of the imperial and colonial character of its Tsarist predecessor. Then, as now, the principal reference point is the relationship of the various non-Russian nations to the Russian nation. As noted earlier, the fundamental line of demarcation is of religious origin. Today, three levels of hierarchy exist within the Soviet Union. Within the Christian group of nations, the Ukrainians and Byelorussians are progressively being integrated with the Russians into a super East Slavic nation and status distinctions among the three have virtually disappeared. The other Christian nations (Estonians, Latvians, Lithuanians, Moldavians, Georgians, Armenians) enjoy an intermediate imperial status whereas Moslem nations are still treated primarily as colonies; however, elites from both the other Christian and Moslem nations are selectively co-opted into the Russian component. To some degree then those elites enjoy an imperial relationship with the ruling Russian nation even if their nation does not.

The Outer Soviet Empire. The outer empire also consists of a series of concentric constellations of states, which are formally independent and participate as full members in the general interstate system, and are members of the United Nations and its affiliated organizations and associations. Some of these states are also integrated into purely Communist political, economic, and military multilateral organizations; others are Third World countries which belong to the Group of 77, the Nonaligned Group of States, and other Third World multilateral bodies, and thus belong to both the Communist world and the Third World.

The innermost ring of the outer empire consists of members of the socialist community/commonwealth. Within this group, the members most tightly integrated into the Soviet imperial structure are the Warsaw Pact states of Eastern Europe, of which the Northern Tier states are the most important and valuable to Moscow (East Germany, Poland, and Czechoslovakia). The Southern Tier is made up of Hungary, Romania, and Bulgaria, of which Bulgaria is closest to Moscow and Romania the most distant.

Albania and Yugoslavia are independent Communist states whose ideological deviation from Moscow is in opposite directions. Neither is a functioning member of the Soviet empire; both are alienated former members. Yugoslavia participates as an autonomous member of certain Communist interstate organizations, but on a sporadic and not fully integrated basis. In East Asia, Mongolia's relationship to the Soviet Union is similar to that of the Warsaw Pact states, whereas North Korea's relationship to the Soviet Union remains both ambivalent and ambiguous because of the Sino-Soviet conflict.

China, of course, represents a former important member of the Soviet empire, which has since defected. It remains in a condition of alienation, although it continues on good terms with other Communist states. The Soviet Union continues to recognize Albania, Yugoslavia, and China as "socialist states," part of the World Socialist System, but not members of the Socialist Commonwealth.

Cuba and Vietnam represent detached members of the outermost ring of the outer empire, since they do not share a frontier with the Soviet Union or with a friendly Communist state. Both are vulnerable, exposed, and represent valuable Soviet assets subject to retaliation. Both are essentially voluntary members of the Soviet empire and their dependency upon the Soviet Union is a reflection

111

of their security concerns. Cuba relies upon Moscow for protection against the United States and to provide an extended protective umbrella for Cuban revolutionary activities in Latin America and Africa. Vietnam needs Soviet protection against China and to provide protective cover for its local imperialism in Indo-China. Cuba and Vietnam share other characteristics in that they are the only other Communist states with their own mini-empires: Cuba with Grenada before October 1983 and Nicaragua; Vietnam with Laos and Cambodia.

Both Cuba and Vietnam are susceptible to what Huntington describes as "weaning" strategies, but the price is apt to be exorbitant. Both are also subject to retaliatory strategies, particularly Cuba, whose existence and survival depends more upon US self-restraint than Soviet protection. Vietnam is not as vulnerable to US power, but more exposed to Chinese power. But, as the Vietnam war demonstrated, punishing Vietnam can be inordinately expensive. Nevertheless, Vietnam and its two dependencies could be fatally vulnerable to joint Chinese-American strategies, although that appears remote at this time.

The Extended Soviet Empire. The cement that bonds the various parts of the Soviet empire together is considerably diversified and varies from one ring to another and from one state or nation to another within each ring. Moscow also has its own priorities with respect to the various segments of its empire. Since its energies and resources are limited, the intensity of its commitments and obligations is accordingly uneven, and Moscow is ready to run varying degrees of risks and to expend varying amounts of costs to extend or protect its empire. Some of its most valuable assets are also the most difficult to bond to the empire, especially the states of Eastern Europe, three of which have defied Soviet control and triggered severe countermeasures (Hungary, Czechoslovakia, and Poland) and one (Rumania) which continues to register defiance. Two (Albania and Yugoslavia) have long since defected, leaving only Bulgaria as the most reliable and self-bonding asset in Eastern Europe. East Germany remains a giant question mark, with the regime obsequiously loyal and the population longing for reunification with West Germany.

Moscow uses diverse instruments to keep its empire together: military occupation, intimidation, bribery in the form of economic and financial assistance, legal obligations, economic and military dependency, and the creation of domestic ideological and political elite constituencies whose fate is inseparable from the Soviet connection.

It would be a mistake, however, to assume that the bonds which hold the empire together are entirely coercive in character. To be sure the closer and more valuable the assets of empire (border non-Russian nationalities and East European states like Poland whose security relevance to Moscow is virtually indistinguishable from Ukraine or Lithuania), the more coercive the instruments that bind. Correspondingly, the more remote the components of the empire, the more voluntary the bonds that connect with Moscow, but also the less valuable the assets.

This observation characterizes the various rings of the Soviet extended empire, including that motley belt of socialist-oriented regimes scattered across the Third World which, according to Moscow's latest count, number more than twenty states with a combined population of 220 million. These self-declared Marxist-Leninist regimes which include among their ranks some of the most underdeveloped societies in the world, barely possess a social structure, and perenially subsist on the margins of existence. Their regimes are as unstable as their societies are volatile, and even Soviet ideologies find it difficult to classify them in accordance with Marxist-Leninist analytical norms.

All of these states represent voluntary adhesions to the Soviet empire and exemplify not so much client or dependent states as supplicants and ideological parasites, whose ideological loyalties in large measure are designed to elicit tangible rewards from the Soviet Union. Their value to the Soviet Union, in many instances, remains dubious. However, their very existence in such profusion and the spontaneity of their association with the Soviet empire reinforces the USSR's global credentials and serves to convert a territorial empire into an overseas global imperial structure, ramshackle and rickety though it may be for the time being.

It is not always easy to identify these states with precision and the concept "socialist-oriented regimes" is improvisational and elastic in its dimensions. Even the ideological affinity of the regimes to the Soviet Union is neither uniform nor clear and many Soviet writers and ideologists are suspicious of the Marxist-Leninist credentials they display and their knowledge and commitment to the ideology. The countries most frequently mentioned in this category are the following: Afghanistan, Laos, Cambodia, South Yemen, Ethiopia, Mozambique, Angola, Congo (Brazzaville), Benin, and Guinea. Sometimes, Tanzania, Zimbabwe, Malagasy, Grenada, and Nicaragua are also included.

These states are characterized by a self-declared adherence to Marxism-Leninism, single party rule under what Moscow calls a "Vanguard Party of Workers" modeled on Communist Party organization structures and procedures, and ruling elites whose domestic social support is often meager or nonexistent and which seek an external constituency as a surrogate. Moscow has lost little time in recognizing the possibilities of reinforcing these local elites, strengthening the bonds of dependency with the Soviet Union, and converting a voluntary dependency into an involuntary one, but one which can also expect only marginal commitment from the Soviet Union in return. Thus, one Soviet writer notes in this connection that the path to follow is the traditional colonial pattern of creating local elites whose fate and fortunes are tied to the colonial power:

> The Socialist states should also give attention to assisting the developing countries in training national cadres—this is one way of exerting a positive influence on these countries. Every year 70,000 to 80,000 citizens of these states are enrolled in various forms of vocational-technical secondary and higher education in the Soviet Union alone.[5]

Soviet observers, furthermore, are not entirely oblivious to the economic possibilities of empire and appear ready to replace Western countries as the chief consumers of the raw materials and resources these countries have to offer. Some Soviet writers are more candid in this regard than is usual for Soviet sources, in anticipation of a restructuring of the international division of labor to Soviet advantage.

> On the one hand, given the international division of labor that is developing within the World socialist system and the food problem that exists in a number of socialist countries, there is another way for the young independent socialist-oriented states to eliminate their backwardness: by intensifying their traditional agricultural production and obtaining machinery and equipment through cooperative arrangements with the developed socialist countries. Of course, the development of economic ties between the socialist countries and the socialist-oriented liberated states is a complex process that entails, among other things, the creation of a certain degree of complementariness between the two groups' economies.[6]

114

And, in return, given the limits of Soviet resources, the Soviet Union increasingly in its economic and military assistance programs displays a preference precisely for those kinds of regimes that show high promise of remaining loyal to the Soviet empire once they join. The Soviet assumption appears to be, not entirely unfounded, that Third World countries whose internal political, economic and social values and norms most closely resemble their own, and are distant from those of capitalist countries, are more likely to remain tied to the Soviet Bloc. Earlier, when the Soviet Union indiscriminately dispensed assistance to Third World countries on a purely opportunistic basis, the results were disappointing. Hence, according to the Soviet authority already cited above:

> In deciding to provide direct political and moral support and material assistance, the Soviet Union and other socialist countries give preference to countries that are carrying out progressive socio-political, social and economic transformations, have embarked on the non-capitalist path of development, and have a socialist orientation.[7]

Furthermore, Soviet observers also rely upon the assumption that an anticapitalist, anti-Western posture by these states will make it difficult to develop a counter-elite that would look towards the West for support. Moscow, it should be noted, is fully cognizant of the internal instability of these regimes and the differing perspectives of what it calls the narrow bureaucratic and professional elites, including the military, and the highly subjective and personality-oriented character of politics, particularly in Africa.

The voluntary character of the bond that unites various states and nations to Russia and the Soviet Union varies considerably. The most reliable voluntary bonds are structural in character, reflecting a deep common security interest, outlook or culture with the Russian nation. Within the Soviet Union, bonds of this character are exhibited by nations like Georgia, Armenia, Byelorussia, the Ukraine and Tadzhikstan. Outside the Soviet Union, bonds of a similar character are reflected by Bulgaria, Mongolia, Vietnam, and Cuba. The socialist-oriented states are tied to the Soviet Union by more fragile opportunistic perceptions, but given the character of their systems, an opportunism that can be replaced with more durable bonds as discussed above.

Another group of states whose orbit around Moscow is furthest removed of all are non-Socialist client states whose connection with the Soviet empire is purely opportunistic and frequently episodic. Currently, Syria is the best example of this kind of dependency, although India also falls to some degree into this category. Originally a number of Third World countries (Egypt, Indonesia, Iraq, Ghana, Mali, and Algeria, among others) also fit into this category and easily fell out of the Soviet orbit once the Soviet connection was no longer of value. These states, unlike the socialist-oriented regimes, carefully excluded Soviet penetrations into their domestic structures, and outlawed and even imprisoned, exiled or executed local Communist leaders. The formal basis of their relationship with the Soviet Union was the ubiquitous Friendship and Mutual Assistance Treaty which obligated Moscow to deliver military, economic and technical assistance in return for vague statements of political and ideological unity.

As noted earlier, as a result of its earlier experiences, the Soviet Union now concentrates on aiding socialist-oriented states, whose regimes are more congenial to Soviet penetration of their domestic social and political structures. Furthermore, in accordance with Soviet ideological predispositions, future leaders in these countries are selected on the basis of socioeconomic criteria which, from the Soviet point of view, would enhance their future reliability:

> It is primarily members of non-privileged families of the workers and the middle social strata who go to socialist countries. Therefore the overwhelming majority of Afro-Asian students in the socialist countries come from families of persons doing work for hire.[1]

VULNERABILITIES AND STRENGTHS OF THE SOVIET EMPIRE

All empires expand and contract simultaneously or sequentially while in the process of expansion, organization, and consolidation. Disintegration and decay set in when empire constuction has exhausted its expansionist energy and assumes a static or terminal condition. A static empire is susceptible to internal ferment and peripheral resistance, which together provide both the centripetal and centrifugal forces which rend the empire apart. The Soviet empire is still in a condition of dynamism and this dynamism, as in the case of previous empires, often generates sufficient energy and incentive to contain internal ferment and resistance. Hence, as long as the Soviet empire is perceived by its rulers and subjects as in an expansionist mode, the chances of internal collapse or explosion are minimized. Only if Soviet expansionism is arrested are internal intractable problems likely to overtake it. The history of empires suggests that once expansion halts, it becomes increasingly difficult to find both the ethical justification for its maintenance and to recruit successive members of a ruling imperialist class whose legitimacy to rule and whose confidence born of success are self-evident. A successfully expanding and dynamic empire is less susceptible to the exploitation of its vulnerabilities than a static one, hence the most immediate objective is to arrest and contain its expansion.

Each of the various concentric components of the Soviet empire has its particular strengths and vulnerabilities. Since these concentric circles, like a Russian Matroshka doll, simultaneously protect and conceal the innermost doll, the most vulnerable but least valuable of its imperial assets are in the outer most rings. There is an old Russian folk-proverb that "the cloth unravels at its edge," and in the case of the Soviet empire, the unraveling process is most easily initiated at its margins.

The core of the Russian empire, the inner empire itself and its East European extension, are correspondingly the most valuable imperial assets and also the most difficult to exploit, since the risks of military conflict and even nuclear war are omnipresent. Moscow has demonstrated a number of times that the region is sufficiently important to use direct military force (Hungary, Czechoslovakia) or its imminent threat (Poland) even though this risks a broader conflict. In late 1979, the Soviet Union for the first time even employed military force in dealing with a part of its extended empire in the Third World, Afghanistan. The fact that Afghanistan bordered on a sensitive area of the Soviet Union, and that Afghan internal disorder took place during a period of uncertain international crisis (the Iranian revolution) when the United States appeared particularly debilitated, played a role in the Soviet decision. Nevertheless, for all practical purposes, Afghanistan is now perceived by Moscow as part of its ring of the outer empire along with Eastern Europe and Mongolia.

Vulnerabilities and Strengths of the Inner Empire. The Soviet inner empire has many vulnerabilities that are essentially chronic in character rather than acute, and are susceptible more to management than terminal resolution. In the long run, these are probably the most serious vulnerabilities facing Moscow, but they pose little possibility of immediate exploitation as part of a deterrent strategy. Nevertheless, under conditions of crisis and disorder, which any war can provoke, some of these chronic problems may become acute and hence more relevant to a defensive strategy. Whether these problems are described as economic, demographic, geographic, regional, military, or labor force skills and distribution, among others, they all are intertwined into a single massive ethnodemographic problem for which the above are principally euphemisms. Uneven rates and levels of population reproduction, births, deaths, urbanization, migration, education, and economic development feed back to create economic, social, military, and political problems. Thus, decreasing rates of reproduction among the ruling Slavic nationalities pose a long-range threat to the cohesion of the empire, since the main impetus for creating, maintaining, and extending the empire comes from the East Slavic, particularly Russian element. And if the inner empire starts to decay, the outer rings of the empire will crumble accordingly. Corresponding to the decreasing growth rates of the Slavic population are the increasing growth rates of the non-Russian nationalities, particularly the Moslems of Central Asia.

According to projections based upon Soviet sources, the ethnographic maldistribution will already become extremely serious by the year 2000 (see Table 6-1). The Soviet population is projected to grow from 262.4 million in 1979 to about 300 million in the year 2000, but of the 38 million new citizens, less than 3 million will be Russians and more than 20 million will be Moslems. Russians, as a percentage of total Soviet population, will have dropped from 52.4 percent in 1979 to 46.7 percent in 2000, and these ethnographic trends will continue into the twenty-first century.

These asymmetrical ethnodemographic dynamics will seriously affect the economy and the military. An increasingly larger share of the incremental growth in the labor force will be drawn from the Moslem nationalties as will be an increasingly larger proportion of military recruits (see Table 6-2).

118

Table 6-1.
SOVIET POPULATION BY ETHNOREGIONAL GROUPS
IN 1979 AND PROJECTIONS FOR THE YEAR 2000
IN MILLIONS AND PERCENTAGES

Group	1979		2000	
	Totals	Percent	Totals	Percent
USSR	262.4	100.0	300	100.0
Slavs	189.2	72.0	195	65.0
(Russians)	(137.4)	(52.4)	(140)	(46.7)
Baltic	5.3	2.0	6	2.0
Armenians and				
Georgians	7.7	2.9	9	3.0
Moslems	43.8	16.7	64	21.3
Others	16.4	6.3	26	8.7

Source: Murray Feshbach, *The Soviet Union: Population Trends and Dilemmas.* (Washington, D.C.: Population Reference Bureau, 1982), p. 22.

Table 6-2.
PROJECTED POPULATION OF WORKING AGES AND MALES AGED 18 IN THE SOVIET UNION ACCORDING TO MAJOR ETHNOGRAPHIC REGIONS

Regions	Working Age Population (in millions)			Males aged 18 (in thousands)		
	1980	2000	Net Increase	1980	2000	Net Increase
USSR	154.8	171.0	16.2	2,238	2,544	306
Slavic	118.8	118.7	-.1	1,251	1,124	127
(R.S.F.S.R.)	83.8	83.5	-.3			
Baltic	4.8	4.8	.0			
Caucasus	8.0	10.8	2.8	170	176	6
Kazakhstan and Central Asia	21.4	34.4	13.0	495	668	173
Moldavia	2.3	2.7	.4			

Source: Feshbach, *The Soviet Union: Population Trends.* pp. 27, 29.

Since the data for Table 6-2 is by Republics rather than nationalities, exact correspondence between the two tables is imprecise. The presence of some 9.3 million Russians and 1.2 Ukrainians in Kazakhstan and Soviet Central Asia and some 24 million non-Russians living in the RSFSR compensates for the distortions somewhat. Nevertheless, the data quite clearly demonstrates that by the year 2000, the percentage of 18-year-old males from the RSFSR will drop to 44 percent whereas the proportionate share of 18-year-old males from the Caucasus, Kazakhstan, and Central Asia will rise to 33 percent of the total population.

The increasing number of non-Russian, particularly Moslem, recruits going into the military will serve to accentuate existing ethnosocial cleavages in the Soviet military between the overwhelmingly East Slavic (mainly Russian) officer corps and the increasing proportions of Moslems, who traditionally have been assigned to construction units. Since many of these Moslem and other non-Russian recruits have a poor command of Russian, increasing complaints have been voiced by the highest military officials concerning the problems this creates for the military as it is faced with the prospect of dealing with a more technologically complicated and sophisticated weaponry. Aside from deficiencies in Russian competency, most Moslem recruits, particularly from Central Asia, are drawn from the villages, where educational facilities and opportunities are severely circumscribed and generally inferior in quality. Thus, Marshal Nikolai Ogarkov, the Soviet Chief of Staff, recently registered an observation which is by no means rare in Soviet writings on population and military issues:

> Considering the question of preparing youth for military service, one especially ought to point out the importance to have a good knowledge of the Russian language. Regrettably, a number of young people will come into the army today with a weak knowledge of the Russian language, which seriously hinders their military training. In the armed forces, as is known, all regulations, instructions, training aids, technical and weapons manuals are in the Russian language. Orders, directives, and commands are also rendered in the Russian language. It is completely understandable that if young people have a weak grasp of the Russian language, it will be more difficult to master weapons and technology entrusted to them; coordination of crews, teams will take place more slowly; and all this in some degree will impact negatively on the combat readiness of subunits.[9]

Another Soviet source makes a particular point of the fact that deficiency in Russian is a problem that is serious among young non-Russians and not only among older and middle-aged non-Russians, where one would expect this:

> Ethnolinguistic changes are an important aspect of current ethnic processes in our country. According to the 1979 census, more than 60% of the USSR's non-Russian population is fluent in Russian—which means 40% is not fluent. We can't help noticing that in many instances the number of pupils in Russian-language schools is lagging behind the rate of increase in the total number of pupils. In several republics, the young people's knowledge of Russian is not as good as that of the middle-aged population. [10]

The situation is further complicated by the fact that the Soviet leadership, because of political and reliability concerns, refuses to use many highly skilled and educated non-Russian nationalities as substitutes for the declining Slavic populations. Thus, many Baltic, German and Jewish recruits, among the most educated and skilled and most appropriately equipped to function in a high-tech military, often find themselves assigned to construction units and other military activities where their capabilities are underemployed. The only other major non-Slavic groups with concentrations of highly educated and skilled populations whose loyalty and reliability can be counted upon are the Georgians and Armenians, who unsurprisingly make up a large proportion of professional military officers who are non-Slavic in origin (see Table 6-3).

The ethnodemographic dynamics and asymmetry which characterize Soviet society also result in the increasing segregation of nationalities into urban and rural nationalities and corresponding ethnosocial cleavages. Thus, even in the major cities of Central Asia (Tashkent, Alma Ata, and Frunze), as well as other industrialized areas in the region, the populations are overwhelmingly Russian (along with a smaller proportion of other Europeans), whereas the countryside remains almost the exclusive domain of the indigenous nationalities. This continues to preserve the essentially colonialist character of the Central Asian Republics, with the ruling nationalities in the cities and in charge of the commanding heights of the economy (see Table 6-4).

The Russian presence in the various non-Russian Republics is another indication of the intensity of the imperial or colonial relationship that characterizes the status of the non-Russian

Table 6-3.
ETHNIC COMPOSITION OF SOVIET SCIENTIFIC
AND SCHOLARLY PERSONNEL, 1974

Nationality	Total Scientists and Scholars		Advanced Degrees			
			Doctorates		Cand. of Science	
	Number	Per-cent	Number	Per-cent	Number	Per-cent
TOTAL	1,108,268	100	29,806	100	288,261	100
Russians	739,522	66.7	16,603	55.7	172,014	60.0
Ukrainians	120,373	10.9	2,901	9.7	33,312	11.5
Armenians	23,873	2.2	1,095	3.7	7,154	2.4
Byelorussians	23,095	2.1	516	1.7	6,332	2.2
Georgians	21,270	1.9	1,177	3.9	6,685	2.3
Azerbaidzhanis	15,609	1.4	755	2.5	5,967	2.1
Uzbeks	14,330	1.3	409	1.4	5,405	1.9
Kazakhs	9,886	.9	253	.8	3,325	1.2
Lithuanians	9,794	.9	224	.8	3,511	1.2
Latvians	6,812	.6	167	.6	2,111	.7
Estonians	5,354	.5	206	.7	1,983	.7
Moldavians	2,919	.3	75	.3	1,208	.4
Tadzhiks	2,886	.3	93	.3	1,051	.4
Kirgiz	2,373	.2	65	.2	856	.3
Turkmen	2,371	.2	57	.2	1,036	.4
Jews (1971)	66,973	6.7	—	—	—	—

Source: *Vestnik Statistiki,* 1974 (no. 4), p. 92.

Table 6-4.
CLASS COMPOSITION OF SOVIET NATIONALITIES, 1970 AND 1979
(IN PERCENTAGES)

Nationality	1970			1979		
	Workers	Office Employees	Collective Farmers	Workers	Office Employees	Collective Farmers
Russians	63	25	12	63	31	6
Armenians	60	25	15	62	31	7
Kazakhs*	65	22	13	64	28	8
Estonians	57	25	18	57	32	11
Latvians	54	23	23	58	28	14
Lithuanians	52	18	30	56	27	17
Byelorussians	53	15	32	59	23	18
Georgians	41	26	33	49	32	19
Azerbaidzhanis	50	21	29	58	23	19
Ukrainians	47	16	37	56	23	21
Kirgiz	41	15	44	56	20	24
Tadzniks	37	15	48	55	15	30
Moldavians	32	7	61	54	15	31
Uzbeks	39	16	45	50	18	32
Turkmens	32	17	51	39	16	45
USSR	57	23	20	60	25	15

Source: Yu. V. Arutunian, "Fundamental Changes in the Social Composition of Soviet Nations," *Sotsiologicheskiye Issledovania* (no. 4), 1982.

*It should be noted that the Soviet definition of "worker" includes farmers on state farms (*Sovkhozy*) and this accounts for the exceptionally large proportion of Kazakhs classified as workers (65 percent), since almost all agriculture in the Kazakh Republic is organized into State Farms because of the Virgin Lands program. Furthermore, about 50 percent of all Russians in Central Asia live in the Kazakh Republic with Kazakhs making up a minority in their own Republic.

nationalities (see Table 6-5). The Russian presence in the Republics ranges from a low of 2.3 percent in Armenia to 40.8 percent in Kazakhstan and 32.8 percent in Latvia. The Kazakhs and Kirgiz are a minority in their own Republics due largely to the enormous proportion of Russians, whereas the Latvians have been reduced to less than 54 percent in their own Republic. The proportion of Russians has been reduced primarily in those Republics where the population growth of the native people is unusually high, although there has also been some out-migration of Russians.

Three principal clusters of nationalities emerge as the locus of greatest resentment of Russian rule and potential resistance: the Baltic nationalities; the Moslem Turkic nationalities in Central Asia; and, the Moslem nationalities of the North Caucasus and Azerbaidzhan. Of these, the most serious potential problem is Central Asia because of strong feelings of cultural and religious oppression and deprivation, where national identity is closely integrated with Islam. Furthermore, the rapid growth of these nationalities, their isolation in the villages, their poorer education levels, their restricted opportunities, and their adjacency to neighboring independent Moslem states where Islamic fundamentalism is flourishing—all contribute to a growing national assertiveness.

The three Baltic nationalities present a different problem. The Balts consider themselves culturally superior to the Russians (many of whom agree) and resent their relatively recent resubordination to Russian rule. Heavily influenced by German and Swedish culture, they consider themselves part of the West, but because of their small numbers (barely over 5 million), they pose no immediate threat to the Soviet empire. The long-term threat which the Balts present to the Soviets lies in the strategic location of their territory on the Baltic littoral and the almost certain expectation that substantial sectors of the population would welcome an invading force, if one could realistically materialize. The large influx of Russians into the area contains any immediate threat by counterbalancing indigenous unreliability with loyal Russians.

Similarly, the North Caucasian nationalities are small in combined numbers and, while the intensity of their resentment against the Russians is deep, their location in the interior of the empire diminishes its impact. However, during World War II, nearly a half dozen of these nationalities were deported *en masse* to

Table 6-5.
INDIGENOUS AND RUSSIAN NATIONALITIES
IN THE UNION REPUBLIC
(PERCENTAGES)

Republic	Native Nationality		Russians	
	1959	1979	1959	1979
Russian	83.3	82.6	*16.4	*17.3
Ukraine	76.8	73.5	16.9	21.1
Byelorussian	81.1	79.4	8.2	11.9
Uzbek	62.1	68.7	13.5	10.8
Kazakh	30.0	36.0	42.7	40.8
Georgian	64.3	74.4	10.1	7.4
Azerbaidzhani	67.5	78.1	13.6	7.6
Lithuanian	79.3	80.1	8.5	8.9
Moldavian	65.4	63.9	10.2	12.8
Latvian	62.0	53.7	26.6	32.8
Kirgiz	40.5	47.9	30.2	25.9
Tadzhik	53.1	58.8	13.3	10.4
Armenian	88.0	89.7	3.2	2.3
Turkmen	60.9	68.4	17.3	12.6
Estonian	74.6	64.7	20.1	27.9

*Non-Russians in the R.S.F.S.R.

Source: L.L. Rybakovsky, "The Interaction of Migrational and Ethnic Process," *Sotsiologicheskiye Issledovania* (No. 4), 1982.

Central Asia and their autonomous units erased because they allegedly cooperated with the Germans. Although they have since been "rehabilitated," anti-Russian feeling continues to run deep.

The Russians have been more successful in Azerbaidzhan, where anti-Russian sentiment has been considerably diminished and the Republic is beginning to appreciate the benefits of Soviet rule and the modernization it has brought.

At one time, the Ukraine was a hotbed of anti-Russian sentiment, but probably more anti-Soviet than anti-Russian, because it suffered disproportionately during the collectivization drive under Stalin, who also bore strong anti-Ukrainian prejudices. Since Stalin's death, with the policies of Khrushchev and Brezhnev (both from Ukrainian border areas), Ukrainian alienation has weakened considerably. The Ukrainians now are officially *secundus inter pares* among the nationalities and represent in effect the junior member of the developing Slavic consortium of ruling nationalities.

As the Russian proportion of the population decreases, the key to maintaining the empire is the *de facto* creation of a super East Slavic nation by fusing the Russians, Ukrainians, and Byelorussians together as the ruling nationalities of the empire. The three Slavic groups account for 72 percent of the total population and will still account for 65 percent in the year 2000. If a bare majority of Russians were able to hold the empire together under great adversity, the combined Slavic group should be able to sustain its rule in the absence of unusual calamities. Khrushchev inaugurated a similar formula in Central Asia by recognizing the Uzbeks as the most powerful and dangerous nationality in the region by reshaping the Uzbek Republic into the Soviet centerpiece of Central Asia. Nevertheless, in spite of both symbolic and substantive gestures, Soviet policies have not been as successful among the Uzbeks as among the Ukrainians and Azerbaidzhanis.

One final point concerning the Ukraine. A small pocket of anti-Russian sentiment exists in the Western Ukraine, the former Eastern Galicia, which was governed from Warsaw between 1919 and 1939 and from Vienna before that. The area was incorporated into the USSR in 1939 and had never been a part of the Russian empire. Its inhabitants were primarily Uniate Catholics and to some degree Polonized. Their long separation from their Eastern Ukrainian kinsmen made them almost into a separate nationality, whose historical and cultural affiliations with Russia were relatively meager.

On the other hand, in spite of the diminution of status and priority since Stalin's death, and subsequent partial alienation, Georgia and Armenia remain among the most reliable, loyal and appreciative, but not submissive, nationalities with respect to the Russian and Soviet connection. Both nations owe their national survival and identity to their association with the Russians who protected them from their Moslem neighbors to the North, East, and South. For the Georgians and Armenians, the term "Elder Brother" as applied to the Russians in terms of protection is widely recognized. Both nationalities remain highly nationalistic and demonstrate their refusal to convert appreciation into submissiveness in many ways. Under Stalin, both nationalities were shown preferential treatment (as collectivities; not necessarily as individuals), especially Georgia, and both resent to some degree that their loyalty and reliability are taken for granted.

Vulnerabilities and Strengths in the Outer and Extended Empires. Whereas the inner empire is an integral juridical component of the USSR bound to the Russian nation by history, law, military force, social, cultural, political and economic integration, ideological uniformity and conformity, the outer and extended empire is juridically separate from the Soviet state and the Russian nation. Various degrees of client relationships, ranging from satellite status (mainly historical) to affiliation based upon opportunistic and episodic common interests (Syria), exist between the USSR and its outer and extended empires. The outermost ring of the Soviet extended empire (opportunistic ring) allows entry on the basis of mutual agreement and exit by unilateral action by either the Soviet Union or its client (Egypt, Somalia, Iraq, etc.).

The members of the extended empire are the most exposed and vulnerable assets of the Soviet Union, subject to both "weaning" and "weakening" strategies, as well as direct military action. Remote geographically, grossly underdeveloped, with fragile social structures and unstable political systems, often adjacent to hostile or unfriendly states, and subject to internal strife and regime transformations, they are at the same time the least valuable to Moscow, whose obligation and commitment to their survival is marginal. Their value to Moscow currently is largely symbolic, enhancing the global credentials of the Soviet state and the universalism of Marxist-Leninist ideology and the Soviet brand of social development. Some countries—Angola, Ethiopia, and South Yemen—have important geostrategic (and some resource) value, and the Soviet investment, but not necessarily commitment, has been substantial. Many of these socialist-oriented countries, however, are socioeconomic basket cases and currently are a drain upon Moscow's resources rather than providing something of tangible value. On the other hand, many are near or adjacent to more important resource-laden Third World countries, particularly in Africa and, hence, while unimportant in themselves, constitute strategic bridgeheads and outposts of a potentially larger and more valuable empire, economically as well as otherwise. Whether such an enhanced extended empire materializes will continue to depend upon the absolute and relative growth of Soviet power and the changing "correlation of forces" which will create both opportunities for expansion and the means for their exploitation.

The innermost ring of the outer empire, on the other hand, is made up of components whose security relevance for Moscow is virtually indistinguishable from the inner empire. This ring is made up of independent states which are full members of the interstate system and have separate contacts with the outside world through diplomatic institutions, treaties, commercial relations, etc. Yet, in essentials, their relationship to the Soviet Union is more akin to the nations of the inner empire than to states in the outermost ring of the outer empire.

Just as the inner empire forms a protective buffer that cushions and isolates the Russian heartland from its enemies, the inner ring of the outer empire serves as a buffer and cushion which protects the inner empire. Without the inner empire the Russian nation would find great difficulty maintaining itself as an integrated territorial state, given its geographical location and its topographical features. Isolated from the oceans and concentrated in the interior of the Eurasian land mass, the ethnographic territory of the Russian nation is totally and utterly bereft of natural defensive frontiers or even a rational and definitive ethnographic configuration.

Whether a nation that currently numbers nearly 140 million people, the largest in Europe, can exist as a viable territorial state within its ethnographic confines is highly dubious. In any event, neither Russian leaders in the past nor Soviet leaders after them have thought so. This suggests that a Russian empire of some sort is inevitable. One must come to the lamentable conclusion that Russian imperialism and Russian survival are inseparable. What we have in the Russian case is an instance of "defensive imperialism," which means a situation where the defense and survival of one nation is dependent upon the extinction or subordination of other nations. While this condition has characterized a number of nations throughout history and even today, only powerful nations are in a position to deny independence and survival as a state to other nations in order to ensure their own existence. For many of the nations of the inner empire, their independence can only be secured by placing the Russian nation at peril.

The defensive character of Russian imperialism must be recognized in order to develop appropriate strategies to deter Soviet expansion and provide for the effective defense of Western Europe. Eastern Europe is not primarily an economic asset; it is first and foremost a security and strategic asset. There are those who argue that Eastern Europe's value to the Soviet Union as a *glacis* or defense zone has eroded because, with the advent of nuclear weapons and instantaneous methods of delivery, the Soviet Union no longer needs Eastern Europe for its security or defense. This takes a very narrow view of security and defense, one that the Soviet leaders would instantly reject. To understand the invalidity of this view, simply imagine a situation in which NATO forces are lined up against the western frontier of the Soviet Union, rather than where they are currently stationed. Just as Russia needs the non-Russian borderlands for its security, Eastern Europe is necessary for the protection of the inner empire and something beyond is needed to provide security for the Soviet grip on Eastern Europe.

The Soviet concept of security thus becomes infinitely regressive. As the new perimeter expands, it needs protection. Expansionism (imperialism) and security have become inseparable in the Soviet view, and the practice of denying security and independence to others in order to ensure the survival of empire has become virtually an automatic reflex.

Thus, Huntington's assumption that Eastern Europe is a valuable Soviet asset that could be held at risk to deter Soviet attack and to carry out the defense of Western Europe is hypothetically attractive. But Eastern Europe is an asset which Moscow will defend with capabilities sufficient to overcome any means that can be mustered by the Western Alliance.

When one looks at a topographical map of Central Europe, it boggles the imagination to contemplate that, as a blitzkrieg-type Soviet assault crosses the North German plain accompanied probably by deep air and missile strikes in NATO's interior, a conventional NATO military force will conduct its own blitzkrieg across difficult mountainous terrain and occupy Leipzig and Prague to set the stage for a fair exchange of assets in return for terminating the war.

Several things are inadequate about this strategy. We will, for the time being, sweep aside political and financial objections—that the European states will not support such a major change in NATO's strategy and that the United States will simply not pay the political and financial costs to assemble, train, and maintain such a force, although these are by no means the least of the many deficiencies in this option. Assuming that such a powerful military force existed, can one imagine it being used for retaliation in the face of a Soviet military attack and occupation of substantial parts of West Germany, rather than employing such a force to repulse the Soviet attack? And when the NATO retaliatory force reaches Leipzig and Prague, then what does it do? Can it go on any further? Huntington mentions Hungary as another target, which is bizarre since Hungary can only be reached if NATO troops cut across Austria or occupy virtually all of Czechoslovakia! Does Huntington assume that the NATO retaliatory force will just keep rolling along or will it halt and wait for the Russians to give up? More likely the Russians will mobilize their reserve forces, which are greater than those of NATO, and launch a counterattack against what would be by then an exposed salient.

Another assumption which Huntington employs is the unreliability of East German and Czech troops. The validity of the assumption depends upon circumstances. Czech troops are not so much unreliable as they are excessively prudent. Czech troops have not defended themselves as a national force since the Battle of White Mountain in 1620; they did not defend themselves on March 15, 1939; they did not defend themselves when the Warsaw Pact troops invaded Czechoslovakia in 1968. They are not likely to defend themselves against a NATO retaliatory force either; unless it is perceived that a NATO victory would result in a redismemberment of Czechoslovakia in favor of a reunited Germany. Under such conditions, they may renounce past precedent.

The reliability of East German troops will depend almost entirely upon how they perceive the outcome. Undoubtedly, many East Germans may desert during the first days of the war when nothing is certain, but it is more likely that the Soviets will commit the East Germans later after they have assessed the probable outcome. No East German army will misperform if it is believed that the Soviet side will win.

What is more likely to deter a Soviet attack upon the West in terms of nonnuclear capability is an enhanced territorial capability developed by West Germany. Moscow has a keen appreciation of the effectiveness of guerrilla warfare. The prospect of an armed West German partisan force operating behind Soviet lines and in occupied territory is more likely to impress Moscow than a conventional military retaliatory force.

It is true, as many scenarios posit, that under certain circumstances the Soviet side might be able to launch a blitzkrieg operation across Northern Germany and reach the North Sea, bypassing the large cities, in a matter of a few days. But what does it do then? It would be confronted with the same problem as Huntington's proposed conventional retaliatory force. Winning the early battles, even in an impressive manner, is insufficient to win the war.

How does one terminate a war to one's advantage after winning the initial battles? It is the problem of what Moscow does after it reaches the North Sea that contributes to deterrence, along with what might be provoked if it prosecutes the war further when nuclear weapons are still lurking in the background. There is no prospect that the Soviet Union can occupy all of Western Europe with its military forces, and occupying only a small part merely results in an inconclusive result. Instead of pouring resources into a conventional retaliatory force, West German capabilities for partisan warfare, which would complicate even more what the Russians would do after their drive to the North Sea, could reinforce the deterrence effect.

We tend to think that a successful deterrence depends upon a single instrument: nuclear weapons, strong conventional forces, etc. Actually, successful deterrence is achieved through multiple instruments, each contributing its share to the mix of ominous uncertainties and absolute certainties.

The nuclear deterrent will continue to contribute its share even if West European credibility erodes and the chances of the use of nuclear weapons are reduced considerably from the viewpoint in Moscow. As long as Moscow perceives the probability of their use as a visible figure above zero (and this is inescapable as long as nuclear weapons are deployed and in sufficient quantity), the nuclear deterrent will continue to function. Particularly will this be true when a nuclear deterrent is considered in combination with the existence of substantial conventional military forces, an American military presence, threats of retaliation outside Europe against Soviet imperial assets in other parts of the globe (Cuba, Vietnam), and Moscow's self-perception of how to consolidate a limited military victory without running the risk of protracted conventional conflict surrounded by hostile populations.

In any combination of deterrent elements, as long as nuclear weapons are part of the package, the costs of misjudgment, misperception or miscalculation are so catastrophic that deterrence will function. Ordinary risk and cost calculations do not apply; in the use of conventional weapons, a risk of conflict below 50 percent will always be more acceptable than risks of even a fractional amount above zero where the possibility of nuclear retaliation exists.

An enhanced territorial capability for West Germany would probably add more to existing deterrents than a more expensive and even more threatening conventional retaliatory strike force. The Soviet military will be able to handle a NATO conventional strike force in East Germany and Czechoslovakia but would find it more difficult to cope with guerrilla warfare as it attempts to press westward and faces additional military hazards and obstacles. It is this vision of a protracted nonnuclear conventional and partisan war that is more likely to deter the Soviet Union than any other nonnuclear force, save the perception of an overwhelmingly superior NATO conventional force.

SOVIET PERCEPTIONS OF THE EFFECTIVENESS OF WESTERN DETERRENT OPTIONS

In evaluating the relative and comparative effectiveness of

various Western deterrent strategies for Europe, the most important factor in the equation is an accurate assessment of what Soviet leaders perceive as the most effective deterrent strategy. As in many similar instances, this involves the extremely difficult and intricate business of making perceptions about perceptions, i.e., a perception of what the Soviet leaders perceive to be the most effective deterrent. One need not be enveloped more deeply in the almost limitless conceptual and cognitive problems that are involved in the process of analyzing perceptions about perceptions, except to say, ultimately, one must also delve into the problem of how the Soviet Union believes that the West perceives it, and so on *ad infinitum*. Recognizing the problem does not solve it, but simply alerts us to the methodological fragility and provisionality of any assessment, which finally rests upon a combination of informed and intuitive judgment rather than absolute empirical proof.

The first issue to be tackled is to determine *what* is being deterred in Europe, and whether Western perceptions of what is being deterred and Soviet perceptions of what is being deterred correlate with Soviet intentions, or are based upon an assessment of Soviet military capabilities from which a range of possible intentions are inferred or imputed. This is not always an easy matter and misperceptions are always a hazard as a consequence. Granted, it is frequently easier to measure military capabilities and deduce a range of possible options than to decipher intentions, and most often strategies are developed to cope with a range of military options which capabilities will support rather than settle upon a single intent.

Western military deterrent strategies generally focus on deterring a Soviet military invasion of Western Europe. This is usually determined by assessing whether Soviet capabilities are in fact sufficient to execute such an invasion successfully. And as long as Soviet capabilities are of such magnitude to carry out an invasion, that option must be deterred, even if that may not be Soviet intent. Many observers seriously question whether the Soviet Union has ever contemplated a pure, direct military invasion and conquest of Western Europe. Rather, the view is that Soviet military capabilities are developed primarily for national defense, the preservation of its control over Eastern Europe, and to deter, nullify, or repel any attempt to employ force or the threat of force to prevent the Soviet Union from carrying out its foreign policy and ideological objectives by nonmilitary means or a combination of political, social, and military means.

Such an imputation of Soviet intent does not presuppose a purely defensive or benign posture, but assumes that Soviet behavior will be *assertive, expansionist* but not always necessarily *aggressive*. Such a view assumes that Soviet military capabilities at various levels and in different regions are for the purpose of primarily providing a protective umbrella over its policies throughout the globe, in order to deter or repel any attempt to *contain* or *limit* its expansion by military means. This means that the use of Soviet military power as a direct means of conquest is likely to be considerably lower than its use as an auxiliary, reserve or supplementary force in conjunction with other means. Its employment is more apt to be indirect than direct: in the first place to nullify any attempt to contain its expansion and in the second place to discourage US and Western efforts to use military force to thwart or reverse what Soviet leaders refer to as "the social processes of history."

But what are "the social processes of history," which the Soviet leaders wish to unleash and protect in Western Europe? The evolution of Eurocommunism in Western Europe, particularly in Italy and Spain, and the unattractiveness of the Soviet model of progress or the pursuit of progress through revolution in Western Europe, would appear to contravene Soviet presuppositions of latent "social processes of history," which are exploitable in West European countries. But, if the views of some revisionist and Marxist-American historians of the "Cold War" are a guide—and they are—the Soviet leaders may be persuaded that American military power in Europe since World War II was designed precisely to demoralize, dampen, and eventually distort historical processes in Western Europe. From the Soviet point of view, the primary function of NATO was neither to repel a Soviet invasion or even to prepare an attack upon Soviet positions in Eastern Europe (in spite of its propaganda), but has been from the very beginning to preserve the social and political *status quo* in Western Europe and to encourage the resurgence of counterrevolution ferment in Eastern Europe, to which it might extend "assistance."

Thus, the Soviet perception of the role of American power in Europe and NATO is the mirror image of the role of Soviet military power and the Warsaw Alliance: to preserve the social and political *status quo* in Eastern Europe and to encourage the development of domestic revolutionary and progressive tendencies in Western Europe, to which it might extend "assistance." The difference, from the Soviet point of view, is that NATO and American power are employed to thwart and reverse history, whereas Soviet power is designed to unfetter history so that it might pursue its inevitable processes, which allegedly are in congruence with Soviet poltical and ideological objectives.

In the Soviet view, the presence of American military power has been the primary reason for the frustration, enervation, demoralization, and paralysis of the revolutionary forces in Western Europe, even to the point of creating the conditions which deformed and pathologized West European Communist Parties into revolutionary cripples and defectives, i.e., into Eurocommunist Parties. Since there is little possibility that American military power can be *removed* from Western Europe without the risk of nuclear war, Soviet strategy has been directed towards its nullification by developing an overwhelming Soviet military presence in Europe, which would deter the employment of American military power to thwart the latent "historical social processes" that would resurface under the protective umbrella of countervailing Soviet military power. This is the general Soviet theory, and one need not go into the details of precisely what the Soviet conception of "historical social processes" may be, except to note that from the Soviet point of view internal shifts in the British Labour Party and the German Social Democratic Party, the emergence of the "Green Party" in West Germany, the "peace movement," etc., represent the latent repressed revolutionary and progressive tendencies in Western Europe that have been fettered by American military power, and are now being released by growing Soviet military power.

Thus, the aim of Soviet power in Europe is a piece of its overall global military power, to provide a protective umbrella over "revolutionary and historical social processes" against the use of American military power to prevent, contain or even reverse them. The Soviet response to "thwarting" historical processes is thus to "free" them, but the Soviet counterpart to "reversing" historical processes is to "assist" their forward movement. At this point the Soviet role moves from the passive to the active, or what may be interpreted by others as the aggressive mode. The invasions of

Hungary, Czechoslovakia, and Afghanistan, as well as various forms of military assistance to Cuba, Grenada, Angola, and Ethiopia, are all empirical manifestations of the seemingly benign term "fraternal" assistance." There are reasons to assume that under appropriate conditions the Soviet Union will be ready and eager to provide such assistance in Western Europe.

Does all this make a difference in the choice of a deterrent strategy in Europe? Indeed it does. The overall Soviet military buildup must be placed in discrete perspective by distinguishing between the role of Soviet nuclear forces and conventional military forces in Europe and distinguishing between Soviet political and military goals, as well as perceiving the interconnections between nuclear and military forces and between political goals and military means.

The role of Soviet conventional forces in Europe has always envisaged the possibility of providing "fraternal assistance" to unspecified solicitants in Western Europe, particularly West Germany, at some unspecified time in the future. Whereas the role of Soviet nuclear forces is to undermine American power as an instrument of maintaining the sociopolitical *status quo* in Western Europe and to *deter* the "export of counterrevolution" and "foreign intervention" (i.e., US efforts to suppress or reverse radical socioeconomic changes), the role of Soviet conventional forces is to repel internal attempts at "counterrevolution" and outside assistance by coming to the aid of the "forces of social progress" upon "invitation."

Soviet conventional forces in Europe have been envisaged as means to extend "fraternal assistance" to Western Europe since the very end of World War II, similar to the assistance provided in Eastern Europe. One need only to reexamine the exchange of messages between Moscow and Belgrade to recognize the line of continuity in Soviet conceptions of "fraternal assistance" beginning in the postwar period and culminating in the tortured Soviet justifications of their invasion of Afghanistan. There is more than a simple coincidence between what Stalin told Tito in 1948 and what the Soviet Ambassador to Paris told an audience in 1980. In 1948, Stalin forcefully reminded Tito that:

> It is also necessary to emphasize that the services of the French and Italian CPs were not less but greater than those of Yugoslavia. Even though the French and Italian CPs have so far achieved less success than the CPY, this is

not due to any special qualities of the CPY, but mainly because . . . the Soviet army came to the aid of the Yugoslav people . . . and in this way created the conditions which were necessary for the CPY to achieve power. Unfortunately the Soviet army did not and could not render such assistance to the French and Italian CPs.[11]

And, in April 1980, the Soviet Ambassador to France, S. C. Chervonenko, warned a French audience that the Soviet Union could "not permit another Chile" and further stated that any country in any region, anywhere on the globe "has the full right to choose its friends and allies, and if it becomes necessary, to repel with them the threat of a counterrevolution or a foreign intervention."[12] Since Chervonenko was the Soviet Ambassador to Prague who staged and orchestrated the massive invasion of Czechoslovakia in 1968, he was well-suited to give a universal application to the "Brezhnev Doctrine," which heretofore was limited only to the Socialist Commonwealth.

Thus, from the Soviet point of view, Soviet conventional military forces are prepared to provide assistance to deter the threat of counterrevolution or foreign intervention, but not to "export revolution," which would be the Soviet counterpart to a military invasion. Since a Soviet military intervention or invasion of Western Europe appears to be contingent upon domestic developments in West European countries and would involve an internal constituency of some sort to which "fraternal assistance" could be extended, a Western preoccupation with deterring a Soviet direct military invasion may serve to deflect attention from the domestic developments upon which a Soviet military move would hinge.

On the other hand, the Soviet leaders always reserve the right to make their own definitions of an appropriate situation or condition, and past experience has demonstrated that they are far from precise or fastidious in their distinctions. What might be perceived as "fraternal assistance," i.e., Soviet military support to repel counterrevolution by the Soviet leaders may be indistinguishable to Western leaders and NATO planners from a Soviet conventional military attack or invasion of a West European country, *a la* Afghanistan.

An effective Western deterrent, whether it be to deter and successfully repel a Soviet military offensive or a Soviet military intervention to assist a domestic insurrection or internal coup in a

West European country, must take into consideration the political and social variables which the Soviet leaders calculate into their strategy. First and foremost, there must be an effective deterrent dealing with the situation in West Germany as a special case within the Western Alliance. West Germany constitutes a buffer and cushion to all of Western Europe. Except for Norway, Greece, and Turkey, for Soviet troops to move into any NATO country, they must first move across West Germany.

The Federal Republic also constitutes a special political and social target. It is the only NATO country susceptible to a Communist-defined civil war strategy, because Germany is divided into Western and Communist states, each of which can lay claim to the other. Although, up to now, the civil war strategy employed by Communists in East Asia (Korea, Vietnam, and China) has not been even hinted at by Moscow, it remains a potential strategy, whose activation will be determined by the course of internal political developments in West Germany. Any substantial alienation in West Germany with respect to either its membership in the Western Alliance or to its existing sociopolitical structure can create an opening for Moscow, with East Germany as its wedge.

Furthermore, the latent force of German nationalism and growing sentiment for the reconciliation, if not reunification, of the two Germanies, can debilitatingly compete with West Germany's loyalty to the Western Alliance and polarize public sentiment. It should be noted that the East German state, in spite of its advanced developmental status, is defined neither as a "People's Republic" nor a "Socialist Republic," but as a "Democratic Republic," which is uniform with the nomenclature defining North Korea and North Vietnam before the unification of the two Vietnams. The concept "Democratic Republic" is a code term denoting simultaneously national fragmentation and provisionality of its state structure. There is little question that at least one future role for the German Democratic Republic is to play a role in the German arena, similar to that of the Democratic Republic of Vietnam in Vietnam, just as the Democratic People's Republic of Korea plays the same role on the Korean peninsula.

Given the geographical location of other Western countries, Soviet intervention elsewhere, whether political or military, would be difficult to execute without a prior change in the status of West Germany. Even the French under President Mitterand are beginning to recognize that West Germany stands as the only geographical barrier between France and the Soviet army, and that the maintenance of West Germany in the Atlantic Alliance and its defense against Soviet intervention constitutes the first line of defense for France. It may soon be difficult for France to sustain its current ambiguity concerning French participation in the defense of West Germany. Increasingly, Paris may recognize that the French military frontier is no longer on the Rhine but on the Elbe, and furthermore that the French nuclear force may have to be extended to explicitly include the protection of West Germany. As will be developed below, this may create the conditions for yet another Western deterrent strategy in place of those already suggested.

In reviewing the various deterrent strategies which are hypothetically possible, whether they be nuclear or nonnuclear, American-dependent or independent of the United States, it would be useful to evaluate Soviet perceptions of their credibility and effectiveness. Ultimately the success of a Western deterrent strategy will depend upon Soviet perceptions, not alliance or individual country perceptions of credibility or effectiveness. Although a given Western deterrent strategy may simultaneously deter the Soviet Union while undermining the reassurance of individual alliance members, the latter determines only whether the deterrent will be established or can be sustained. Nevertheless, Soviet perceptions of its credibility and effectiveness will be the decisive factors, as long as the deterrent is in place. That is why, even though West Europeans may have increasingly less assurance that the United States will risk its own cities and populations by maintaining its nuclear umbrella over Western Europe even under conditions of nuclear parity, deterrence will hinge on Soviet perceptions of credibility, not European.

The loss of European reassurance thus does not threaten the credibility of the American deterrent as long as it is in place. However, if European actions should threaten the existence of the deterrent itself, then obviously there can be no credibility. That is why an alternative deterrent strategy that would revive reassurance, while sustaining credibility, is so crucial.

Eight alternative Western deterrent strategies are evaluated in terms of both Soviet and West European perceptions. Five are nuclear and three are nonnuclear options. Soviet perceptions of credibility, effectiveness, escalation control, and deterrent value are estimated, whereas West European perceptions of credibility, reassurance, willingness to accept the deployment and/or pay the costs of deployment (acceptability), and deterrent value are rated. These evaluations are those of the author, based upon his best judgment (see Table 6-6).

The eight alternative Western deterrent strategies are as follows:
- US Strategic Force (USSF).
- US Intermediate Nuclear Forces (USINF).
- Enhanced British/French Nuclear Forces (EBFNF).
- All-NATO Nuclear Force (ANNF).
- Independent European Nuclear Forces, including a West German Nuclear Force (IENF).
- Enhanced NATO Conventional Defensive Force (ENCDF).
- NATO Conventional Retaliatory Force (NCRF).
- Enhanced European Territorial Defense (EETD).

The perceptions, which are rated on a scale from 0-10, but expressed in terms of low, medium, and high are as follows:

Credibility: the likelihood of use.

Effectiveness: likelihood of success in repelling or nullifying Soviet military action, if employed.

Escalation Control: degree to which Soviet leaders are confident that escalation can be controlled *horizontally* (geographical) or *vertically* (level of nuclear weapons).

Reassurance: likelihood of deterring or repelling a Soviet attack without widespread death and destruction in Western Europe.

Acceptability: willingness to accept the deployment and/or to pay the costs involved.

All of the deterrent strategies, with the exception of the NATO Conventional Retaliatory Force, have a high deterrent value from the Soviet perspective. The reason for the low deterrent value of the NCRF is that the Soviet planners are better able to contain and repel or nullify such a force and thus the Soviet perception of its effectiveness is low. This is true even though Soviet writers have indicated an enhanced interest not only in US plans to enhance NATO conventional forces, but also the strategy of waging the conventional conflict on the territory of the Warsaw Pact states. Thus, one Soviet commentary observes:

> The Pentagon has latterly been sharply increasing the potential of conventional arms. . . . The latest conventional arms are to supplement the

141

Table 6-6.
SOVIET AND WEST EUROPEAN PERCEPTIONS
OF WESTERN DETERRENT STRATEGIES

Deterrent Strategy	Soviet Perceptions of:					West European Perceptions of:			
	Credibility	Effective- ness	Escalation Control Horizontal	Vertical	Deterrent Value	Credibility	Reassurance	Acceptability	Deterrent Value
USSF	Low (3)	High (10)	Nil (0)	Nil (0)	High (8)	Low (3)	Nil (0)	In Existance (10)	Low (3)
USINF	Medium (7)	High (10)	Medium (6)	Medium (6)	High (10)	Medium (7)	Low (3)	In Progress (9)	Medium (6)
EBFNF	Low (2)	High (8)	High (8)	High (8)	High (8)	Low (2)	Low (4)	Medium (6)	Low (4)
ANNF	Medium (6)	High (8)	High (8)	High (8)	High (9)	High (10)	High (10)	Low (3)	High (10)
IENF	High (9)	High (9)	High (9)	High (9)	High (10)	High (10)	Low (2)	Low (1)	High (10)
ENCDF	High (10)	High (10)	High (10)	High (10)	High (10)	High (10)	High (10)	Low (3)	High (10)
NCRF	High (10)	Low (3)	High (10)	High (10)	Low (3)	High (10)	Medium (6)	Medium (6)	Medium (6)
EETD	High (10)	Medium (7)	High (10)	High (10)	Medium (7)	High (10)	Medium (6)	Medium (6)	Low (4)

These evaluations are in reference to Soviet and West European reactions to a possible Soviet military offensive or intervention in West Germany, which is the scenario most often considered and upon which NATO plans are based.

potential of nuclear weapons. The aim remains the same—to achieve military superiority over the socialist countries. While former plans envisaged operations between the Elbe and the Rhine, now—in any war, nuclear or non-nuclear—it is intended to conduct them on the territory of the Warsaw Treaty countries. The Atlanticists hold that this will reduce destruction and losses among the civilian population in the NATO countries to the minimum [i.e., enhance reassurance].[13]

It is evident from this commentary that the Soviet leaders do not view US and NATO military plans in unidimensional terms, but in multidimensional contexts. Nuclear and nonnuclear forces are perceived as being mutually reenforcing, and the strategy of carrying the war to the territory of the Warsaw Pact states is not defined in terms of a limited and focused retaliatory action, but as part of an integrated nuclear-conventional force offensive. It is, of course, this type of multidimensional offensive deterrent strategy which the Soviet leaders fear most and would find the most effective as a deterrent, but which they view as a nondeterrent offensive strategy as well. Of course, it is precisely this type of capability that the West Europeans would refuse to pay for, and a multidimensional nuclear/conventional offensive capability, while the most effective in deterring Moscow, would simultaneously be the most politically controversial and destabilizing in both Western Europe and the United States. It is precisely the type of strategy and capability which domestic societies, even less an alliance of democratic societies, would find the most difficult to deploy and sustain. But, it is also the type of strategy and capability which the Soviet leaders would like to develop, since it would be the most functional and effective for their purposes.

From the Soviet standpoint, any deterrent strategy involving nuclear weapons, irrespective of the level of credibility, as long as it is above zero, has a high deterrent value, because its effectiveness would be high in terms of nullifying or repelling Soviet military action. Any Soviet military initiative, under existing conditions, that might trigger the use of nuclear weapons by the West would be considered unacceptable because of the inconclusiveness of the outcome and the certainty of vast destruction and massive carnage. That is why it is still imperative that the West not adopt a "no first use of nuclear weapons" policy in the absence of powerful countervailing conventional force. The adoption of such a policy would in effect deprive the West of any effective deterrent strategy, since the two nonnuclear options which have a high deterrent value, the Enhanced NATO Conventional Force and the Enhanced Territorial Defense Force, are far from operational, and their likelihood of acceptance by Western Europe is not high. It is important to note that while West European perceptions of deterrent strategies are important for morale and allied relations, there is little or no correlation between West European and Soviet perceptions of credibility, effectiveness or deterrent value. Thus, although the US Strategic Deterrent strategy has low credibility and reassurance in West European calculations, what impresses Moscow is that it is a force in existence and whose effectiveness if employed is very high and hence whose deterrent value is also high. These high values, together with the horizontal and vertical scale of the escalation, are sufficient to countervail against the relatively low Soviet perception of its likelihood to be employed.

In terms of effectiveness and deterrent value, an independent West German nuclear force would rate extremely high. Moscow would be certain that West Germany would use its own nuclear force in its own defense, but it is also aware that such a deterrent strategy would have high West European credibility and a very low reassurance quotient, which renders it almost as unacceptable to Western Europe as to the Soviet Union. This also serves to explain why the Soviet rating of the effectiveness of the USINF nuclear deterrent is very high. Not only is it a deterrent strategy in the process of actual deployment; moreover, from the Soviet perspective, it is the closest approximation to a West German nuclear force, and indeed can easily be converted into one if the United States decides to turn over control of the 108 Pershing IIs and an equal number of Ground Launched Cruise Missiles to West Germany. Overnight, West Germany could become the second most potent nuclear power on the continent of Europe.

Nevertheless, in summary, it should be emphasized that the most effective practical deterrent strategy for the West remains, not a simple, comprehensive unidimensional deterrent, but a combination of deterrent capabilities and strategies, with the capacity to respond to a variety of situations, in which nuclear weapons continue to play a critical reenforcing, if not always a reassuring, role.

ENDNOTES

1. Michael Howard, "Reassurance and Deterrence," *Foreign Affairs,* vol. 61, no. 2, (Winter 1982/83), pp. 309-324. For a further elaboration of his ideas, see "Deterrence, Consensus and Reassurance in the Defense of Europe," in *Defense and Consensus: The Domestic Aspects of Western Security, Part III,* Adelphi Papers No. 184 (London: International Institute for Strategic Studies, 1983), pp. 17-26.

2. Samuel P. Huntington, "The Renewal of Strategy," in *The Strategic Imperative: New Policies for American Security,* ed. Huntington (Cambridge, Mass.: Ballinger, 1982), pp. 21-22.

3. Samuel P. Huntington, "Broadening the Strategic Focus," in Adelphi Papers No. 184, pp. 27-32.

4. Of course, the military balance as perceived by the Soviet leaders serves to determine not only the value of the assets to be protected, exchanged or surrendered, but also to shape the mode and circumstances of their disposition.

5. Yu. S. Novopashin, "The Influence of Real Socialism on the World Revolutionary Process," *Voprosy Filosofii,* no. 8 (August 1982), pp. 3-16.

6. *Ibid.*

7. *Ibid.*

8. V. F. Li, "The Political Superstructures in Societies of Socialist Orientation," *Voprosy Filosofii,* no. 9 (September 1981). See, also, G. Kim, "The USSR and National-State Construction in Developing Countries," *International Affairs,* no. 1 (January 1983), pp. 35-44.

9. N. V. Ogarkov, *Vsegda v Gotovnosti k Zashchite Otechestva* (Moscow: 1982), p. 64. Another Soviet sociologist further notes: "Instruction in the Russian language is not one of the army's main functions; it has many of its own tasks to fulfill. But because some men do not have sufficient command of Russian, it is sometimes necessary in some military subunits to set up language study circles for a part of the soldiers of non-Russian ethnicity. Of course, this makes army service, with its already stressful, full-time course of study, all the more complex." Yu. V. Arutunian, *et. al.,* eds., *A Preliminary Ethnosociological Study of the Way of Life* as translated in *Soviet Sociology,* vol. 21, no. 4 (Spring 1983), pp. 89-90.

10. Yu. Bromley, "Ethnic Processes in the USSR," *Kommunist,* no. 5 (March 1982), pp. 56-64. The problem varies from one nationality to another. According to the 1979 census, the percentage of non-Russians who are *not* fluent in Russian is as follows: Ukrainians, 50.2; Byelorussians, 43.0; Estonians, 75.8; Latvians, 43.3; Lithuanians, 47.9; Moldavians, 52.6; Georgians, 73.3; Armenians, 61.4; Azerbaidzhanis, 70.5; Turkmen, 74.6; Uzbeks, 50.7; Kazakhs, 47.7; Kirgiz, 70.6; Tadzhiks, 70.4. The figures for Estonians, Uzbeks, and Kazakhs are highly suspect. Many Estonians, as a gesture of anti-Russianism, refuse to concede that they are fluent in Russian. Their nonfluency rate is probably less than 50 percent, similar to Latvians and Lithuanians. The Uzbek and Kazakh figures are unrealistically low and reflect the zealotry of local Party leaders. They are probably in the range of 70 percent, similar to the level for other Moslem nationalities.

11. *The Soviet-Yugoslav Dispute* (London: Royal Institute of International Affairs, 1948), p. 51.

12. As quoted in Flora Lewis, "Kremlin's European Policy," *New York Times,* April 22, 1980, p. 14.

13. R. Simonyan, "Aggression with 'Conventional' Arms," *New Times,* no. 9 (February 1983), p. 22.

CHAPTER 7

POTENTIAL SOVIET RESPONSES TO A NATO RETALIATORY OFFENSIVE STRATEGY

by

Daniel S. Papp

If NATO were to adopt a conventional retaliatory offensive strategy for West European defense, what would the Soviet responses be? While no definitive answer to this straightforward question can be given, the nature of the Soviet response would largely determine the wisdom of adopting such a strategy, for it is the combined effect of a conventional retaliatory offensive strategy and the Soviet responses to that strategy that would determine whether deterrence in Europe and beyond had been strengthened or weakened. Doubtlessly, the USSR would respond politically, diplomatically, and militarily to any change in NATO military strategy. The degree to which the USSR believed that it could politically or diplomatically benefit from its responses, and the degree to which the USSR believed it could militarily frustrate the hypothetical new NATO strategy, would therefore determine the extent to which a NATO conventional retaliatory offensive strategy would enhance—or degrade—deterrence.

Likely Soviet responses to a new NATO military strategy may be usefully divided into two broad categories, diplomatic and political responses on the one hand, and military doctrinal and tactical responses on the other. Diplomatic and political responses may be further categorized into those directed toward the United States and Western Europe; toward Eastern Europe; and toward the Third World, including China. Military doctrinal and tactical responses may be discussed under the rubric of withholding strategies; additional advanced deployment; territorial defense; first echelon enhancement; static defenses; and chemical and nuclear options. Other Soviet responses, such as reduction or elimination of oil and gas exports to Western Europe or accelerated growth of the Soviet defense establishment, are possible but not likely. In the case of oil and gas exports to Western Europe, the USSR relies on energy exports to obtain over half of its hard currency intake, which is needed to pay both for grain and technology imports. In the case of an accelerated growth of the Soviet defense establishment, the Soviet economy already faces serious problems that would only be worsened by further expansion of the estimated current three to ten percent annual growth rate of the Soviet military budget.[1] While these and other responses cannot be dismissed out of hand, neither can they be viewed as probable responses.

This chapter, then, will concentrate its analysis on these nine types of political-diplomatic and military responses as we seek to answer how the USSR would respond to a hypothetical NATO adoption of a retaliatory offensive strategy. We will conclude with an assessment of the impact those combined responses may have on deterrence in Europe and beyond.

DIPLOMATIC AND POLITICAL RESPONSES

If NATO seriously were to consider the adoption of a retaliatory offensive strategy, major intra-NATO disagreements doubtlessly would emerge. Indeed, debate over the new strategy within individual NATO nations would probably be even more heated than the intra-NATO disagreements. This debate, along with accusations about the alleged hostile nature of NATO, most assuredly would spill over into Eastern Europe and probably the Third World as well.

If past practice can be used as a guide, the USSR would launch a massive diplomatic and political offensive against the new strategy. US and West European public opinion—and through public opinion, policy—would be the primary target of this Soviet offensive. East European and Third World public opinion and policy would be secondary targets. Moscow would have four objectives. First, it would want to frustrate NATO's adoption of the new strategy. Second, it would try to splinter or fractionate NATO. Third, Moscow would work to strengthen East European adherence and loyalty to the Warsaw Pact. Finally, the USSR would attempt to degrade the West's image and credibility in the Third World. Moscow's level of success would depend upon existing circumstances and undoubtedly would vary; a key point to remember here, however, is that initiation of a diplomatic-political offensive does not imply that the USSR necessarily would fear the new strategy. Rather, the USSR could view NATO's efforts to institute a new strategy as an opportunity to achieve other Soviet objectives—fractionation of NATO, enhancement of East European loyalty, and degradation of the West's international image.

The Political-Diplomatic Campaign in the United States and Western Europe. Any Soviet political-diplomatic campaign would center on NATO's aggressive intentions and would attempt to prove that the United States sought to launch a war that was not in Western Europe's interest. Such allegations would, of course, be nothing new. Throughout NATO's history, the USSR has charged that the alliance had hostile intent.[2] Similarly, and particularly in recent years, the Kremlin has maintained that alleged American designs for a European war diverge from actual European interests.[3]

The success that the USSR could have in such a campaign is difficult to gauge. When the USSR unleashed its verbal assault during 1977 and 1978 against the proposed deployment of enhanced radiation (ER) weapons ("neutron bombs"), the assault strengthened existing anti-ER sentiment. During that assault, the USSR attempted to influence not only public opinion, but also government opinion. Leonid Brezhnev, for example, sent personal messages to West European heads of state warning them about the dangers that ER deployment would bring.[4] More recently, the Soviet Union undertook an even more strident campaign against the proposed deployment of intermediate range nuclear forces (INF).[5] Again, Soviet anti-INF propaganda sought to influence both public and governmental opinion, and stressed that INF deployment would heighten the danger of war. As in the case of the Kremlin's anti-ER effort, the Soviet criticism bolstered indigenous West European INF opposition; how much it increased the opposition, however, is not clear.

A Soviet political-diplomatic assault in Western Europe in response to a proposed NATO retaliatory offensive strategy doubtlessly also would emphasize the divergence of West European and American interests. In view of the fact that West European and US leaders have recently expressed concern over the so-called "Successor Generation" problem,[6] such an approach might appear particularly attractive to a Soviet leadership bent on undermining the solidarity of NATO. The argument that West European interests lay in Western Europe has been used regularly by West Europeans to withstand US pressures to expand West European presence in the Persian Gulf and the Indian Ocean;[7] the same argument would be used by the USSR in an effort first to split Europe from the United States, and then the European peoples from their governments.

NATO claims that a retaliatory offensive strategy would reduce NATO reliance on nuclear weapons would of course be met with derision by the USSR unless sizable quantities of nuclear weapons were withdrawn from the European theater. Even if this were to occur, Soviet assertions that NATO offensive plans were indicative of Western aggressiveness would overcome at least some of the good will garnered in the US and European antinuclear movements by the hypothetical withdrawal of nuclear weapons.

All things considered, then, the Soviet Union would clearly have a number of separate approaches it could take in a US and West European political-diplomatic campaign against a NATO conventional retaliatory offensive strategy. The overall level of success that the campaign would have, however, is impossible to predict. Given the Western alliance's recent history of disagreements about policy following the capture of American hostages in Iran, the Soviet invasion of Afghanistan, and the declaration of martial law in Poland, and given the continuing uncertainty about curtailing economic links with the USSR, ample opportunity would exist for the USSR to launch a political-diplomatic campaign in Western Europe and the United States against a NATO conventional retaliatory offensive doctrine proposal. That campaign could have a considerable degree of success both in frustrating the adoption of the new strategy and in further fractionating NATO.

The Political-Diplomatic Campaign in Eastern Europe. In Eastern Europe, Soviet propaganda against a NATO retaliatory offensive doctrine proposal would again probably concentrate on the claim that the doctrine proved NATO's hostile intentions. Official East European response would coincide with the Soviet line. However, unlike the campaign in Western Europe, the East European campaign would have as its primary objective the solidification of the Warsaw Pact.

Again, Soviet verbal assaults against NATO with this objective in mind are nothing new. Even before NATO was created, the USSR claimed that the United States and other West European governments sought to overthrow the newly-created socialist regimes in Eastern Europe, and the Kremlin justified its 1956 invasion of Hungary and its 1968 invasion of Czechoslovakia as efforts to prevent counter-revolution sponsored by the West. So, too, during 1980 and 1981, the USSR characterized the growing influence of the independent Polish trade union, Solidarity, as a counter-revolutionary offensive supported if not controlled by the West.[8]

To all but the most dedicated Communist, these Soviet assertions were clearly transparent. The USSR had little if anything to support its claim either of NATO aggressiveness or of concrete Western support for the Hungarian, Czech, and Polish revolutions. In much of Eastern Europe, then, Soviet propaganda against NATO is seen at unofficial levels as disingenuous and false. Despite its considerable efforts to portray NATO and the West as aggressive, the USSR has had limited success, and the USSR continues to be regarded in much of Eastern Europe as a domineering overlord.

NATO adoption of a conventional retaliatory offensive strategy, therefore, may have either of two opposite effects. On the one hand, it could greatly enhance the credibility of Soviet claims of

Western expansionism and hostility. For the first time, the Kremlin would have solid and documentable proof that NATO did, in fact, intend to move east during time of conflict. NATO's assertions that an offensive into Eastern Europe would be launched only in the event of a Warsaw Pact attack on Western Europe may carry some weight in offsetting the probable Soviet assertion, but how much is open to debate. Again, the key point is that for the first time the USSR would have documentable proof that NATO planned to strike east. Any effort on the part of NATO to link its new strategy to events outside of Europe would only present the Soviets with a stronger case to make in Eastern Europe that NATO was, in fact, an aggressive bloc.

On the other hand, for those East Europeans who still hoped to escape Soviet domination, announcement of a NATO retaliatory strategy may offer some hope, and through hope a willingness to oppose Soviet policies and interests. This hope and willingness from a Western perspective can only be seen as a two-edged sword. In the event of worsening tensions in Europe, Soviet uncertainty about its Warsaw Pact allies would be enhanced, and deterrence correspondingly would be increased to the degree Soviet uncertainty were enhanced. Conversely, the Soviet Union and its East European allies may respond to increased uncertainty by returning to Stalinist practices and methods in Eastern Europe. Such a response would neither strengthen deterrence nor be in the interests of those East Europeans who seek a loosening of Soviet control.

The overall impact of a Soviet political-diplomatic campaign in Eastern Europe is likely, therefore, to draw the Warsaw Pact somewhat more closely together, because of an increased elite perception of threat and because of a possible return to more strict practices of societal control. Somewhat paradoxically, such Soviet and elite East European responses may make both East European support for the Warsaw Pact more likely in the event of a Warsaw Pact offensive, and East European support for a NATO retaliatory offensive more probable in the event of a NATO penetration into Eastern Europe. How these opposed East European responses would be factored into Moscow's calculus of deterrence is an open question.

The Political-Diplomatic Campaign in the Third World. Soviet successes in the Third World would depend on previous predilections toward NATO and the Warsaw Pact. Those Third World capitals already favoring the Soviet interpretation of world affairs probably would accept Moscow's version of the rationale for the new strategy, and pro-Western Third World states would accept NATO's version. Those Third World states that were legitimately nonaligned would probably conclude that the debate did not greatly affect them, and would continue their nonaligned policies and perceptions. All things considered, either the proposal or adoption of a retaliatory offensive strategy by NATO would likely have little significant effect on the Third World.

China may be an exception. With the People's Republic of China's (PRC) legitimate concern over the security of its northern border, Beijing is vitally interested in European security questions. This Chinese interest in European security issues dates to the mid-1970s and is closely related to China's awareness that a strong NATO requires the USSR to deploy forces in Europe that may otherwise be arrayed against the PRC.[9] Any significant change in NATO doctrine is therefore of vital interest to Beijing.

The USSR realizes this, and would probably structure its political-diplomatic campaign toward China to emphasize that NATO's contemplated strategy change both endangered Chinese security and was undertaken without discussions with China. The success of this Soviet campaign would depend upon Chinese perceptions of Moscow's ability to blunt the postulated retaliatory offensive without transferring units or equipment from the Sino-Soviet border. Three different campaigns would probably appear plausible to the Soviets. First, Soviet units and/or equipment conceivably could be transferred from the Sino-Soviet border area, in which case the PRC leadership may view NATO's contemplated retaliatory offensive strategy as advantageous. Second the new strategy may have no influence on the military situation in the Far East, in which case the PRC would do little and say little either about the strategy or about Soviet assessments of it. Finally, and rather improbably, the USSR may threaten China with an increased military buildup on the border. This option could well lead to a heightened internal Chinese debate about its policies both toward the USSR and the United States. In no event, however, would Chinese—or Third World—responses to a Soviet political-diplomatic offensive against a proposed NATO retaliatory offensive strategy in Europe have a significant impact on deterrence in Europe.

MILITARY DOCTRINAL AND TACTICAL RESPONSES

If the USSR's political-diplomatic offensive against NATO's proposed new strategy proved ineffective, the USSR and its Warsaw Pact allies would find it necessary to alter its military doctrine and tactics to counter that strategy. (Indeed, even before the strategy were adopted, the Warsaw Pact may somewhat revise its strategy and tactics, if not to add to Western debates about the wisdom of a retaliatory offensive strategy, then to prepare the Warsaw Pact to counter the strategy when adopted.) The degree to which the Warsaw Pact may alter its strategy and tactics would be tied closely to the degree of threat the Warsaw Pact perceived the new strategy to present. That threat would be a function of the Warsaw Pact's perceptions of NATO's capability to execute such a strategy and of Warsaw Pact weaknesses in defending those regions where NATO may retaliate. Six Soviet/Warsaw Pact military doctrinal and tactical responses to a conventional retaliatory offensive appear probable and will be discussed below. This section is based on two important assumptions. First, Moscow will believe that NATO's new strategy presents a serious threat to Warsaw Pact attainment of its objectives during war. Second, a significant across-the-board expansion of Soviet and Warsaw Pact conventional capabilities will not occur.

Withholding Strategy. The USSR currently fields thirty divisions in Eastern Europe, nineteen of which are in the German Democratic Republic (GDR). Most Western analysts speculate that in the event of war in Europe, the Group of Soviet Forces in Germany (GSFG) would be fully committed to the fray in either the first or second echelon attack. Given the uncertainties presented by a NATO retaliatory offensive, the Warsaw Pact's Joint High Command (JHC) may opt to decrease the weight of a first or second echelon attack to the degree to which it believed NATO's retaliatory offensive strategy presented a threat to the Warsaw Pact's rear.

Such an option is exactly what Huntington hopes that his conventional retaliatory strategy would cause Moscow to choose. With a weaker initial thrust and/or weaker follow-on support, the Warsaw Pact's chances of attaining a breakthrough at the front would be lessened. Correspondingly, if NATO forces successfully contained a Warsaw Pact attack, NATO would have no need to employ nuclear weapons, and may even choose *not* to implement its retaliatory offensive strategy. Deterrence would in all probability be strengthened. From almost any perspective, then, Soviet adoption of a withholding strategy appears to be exactly what NATO would hope for. Not surprisingly, if simply because of this reason, the JHC would choose a different alternative.

Additional Advanced Deployment. If a withholding strategy would appear unattractive to JHC planners, deployment of additional Soviet forces in Eastern Europe may appear to offer an acceptable solution to problems caused by a NATO retaliatory offensive strategy. Currently, the USSR deploys seventy-six divisions in its European Military Districts. Some are Category 1 divisions that could be deployed to Eastern Europe on short notice. Others are Category 2 and 3 divisions that would have to be upgraded before advanced deployment could occur.

Additional forward deployments would probably require Moscow to revise or renegotiate its Status of Forces agreements with the German Democratic Republic, Poland, Hungary, and Czechoslovakia. The Political Consultative Committee (PCC) of the Warsaw Pact probably could undertake these revisions with a minimum of delay, particularly if the groundwork for revisions had been laid by an effective Soviet political-diplomatic offensive in Eastern Europe against the NATO retaliatory offensive strategy. By emphasizing NATO's aggressiveness, such an offensive also would have had the salutary effect of reducing anti-Soviet sentiment directed at additional Soviet troops being stationed in Eastern Europe. From the Soviet perspective, the two biggest drawbacks attached to additional forward deployments (beyond any increase in anti-Soviet sentiment the deployment engendered) would be the increased financial costs associated with stationing additional Soviet combat units in Eastern Europe and the increased strain placed on logistics.

Additional units in Eastern Europe could be used to good advantage in several ways by the Soviets. They could be added to existing plans for first or second echelon attack, thereby, from the Soviet perspective, making a rapid attainment of war objectives even more rapid, perhaps even before NATO could unleash its retaliatory offensive. Alternatively, additional forward deployments could be held back and enable the JHC to implement an effective withholding strategy without weakening the first or second echelon attack. As a third choice, more units could be placed into mobile "operational maneuver groups" to maximize Warsaw Pact flexibility. In all three cases, NATO's retaliatory offensive strategy would be threatened.

If the USSR were to respond to a NATO retaliatory offensive strategy by deploying additional forces in Eastern Europe, it is probable that the overall level of deterrence in Europe would be little altered. The USSR would have more troops in Eastern Europe and NATO would be prepared to advance east if Soviet/Warsaw Pact forces struck west, but militarily little else would have changed.

Territorial Defense. As we have already seen, a Soviet political-diplomatic campaign in Eastern Europe that preceded a NATO adoption of a retaliatory offensive strategy would probably emphasize the heightened NATO threat to Eastern Europe and the Warsaw Pact. A logical response to such a threat, and one that would be militarily, if not economically, palatable to non-Soviet Warsaw Pact members, would be to increase East European contributions to East European defense. These contributions could come either in the form of military units assigned to the Warsaw Pact, or as units outside the Warsaw Pact designated for territorial defense.

From an East European perspective, either alternative would necessitate additional military expenditures. Since most East European economies are already hard pressed, some opposition to either alternative could be expected.[10] Of these two choices, the latter appears preferable, not only because costs probably would be less, but also because, in a political sense, national officers would be commanding national forces in defense of the national homeland. A more effective defense, therefore, probably could be mounted against a NATO attack even while the potential for NATO success in undermining the local regime was reduced.

From the Soviet perspective, however, both additional non-Soviet Warsaw Pact units in the Warsaw Pact and territorial defense must appear to have significant disadvantages. Some Soviet uncertainty undoubtedly exists about the loyalty of non-Soviet Warsaw Pact troops to the Pact. This uncertainty would probably continue to exist even if it were persuasively argued that the NATO threat had heightened. Because of this, the USSR may argue for increased East European contributions to the Warsaw

Pact, but would probably oppose territorial defense even if it believed territorial defense could effectively frustrate a NATO retaliatory offensive strategy. Possible uses for territorial defense are just too diverse, including opposition to "socialist solidarity" carried out by Soviet arms under the Brezhnev Doctrine.[11]

First Echelon Enhancement. Soviet military doctrine in its simplest form posits that first echelon armies of a front will attack and penetrate enemy defensive positions. Then, at the appropriate time and locations, the front commander will employ second echelon forces to exploit the advantages achieved by the first echelon.[12] NATO and US studies have concluded that NATO forces, as they are currently configured, can successfully stalemate the initial Warsaw Pact attacks, but may not be able to contain the Pact's follow-on strategic second echelon assault. Thus, the US Army has developed the AirLand Battle (ALB) and Supreme Headquarters Allied Powers Europe (SHAPE) Follow-On Force Attack (FOFA) as two concepts which in one way or another advocate deep attacks—albeit in different ways.[13]

Soviet military doctrine is directed at achieving a quick battlefield victory through the use of mobile, rapid operations. By using such tactics against the NATO forces in contact, the Soviets hope not only for a quick, decisive battle, but also for a quick, decisive war as well. Because of this emphasis, it is vital that the attacking echelons not be stopped or slowed by the defending NATO forces. The Soviets intend to accomplish this by concentrating preponderant combat power at the point of the attack—so preponderant, in fact, that it will overwhelm the NATO defenses quickly. From NATO's point of view, then, it is critical that the forward edge of the battle area (FEBA)—the troops in contact—not collapse before the retaliatory offensive begins.

Soviet planners would realize this, too, and may react by strengthening their first echelon forces from either second echelon forces or units formerly deployed in European Military Districts of the USSR. Their objective in strengthening first echelon forces would be to achieve significant success in the initial attack. From the best of all Soviet perspectives, a strengthened first echelon attack may even present NATO with a *fait accompli*—with defeat imminent, why go east? At worst, a strengthened first echelon attack may force NATO decisionmakers to make the decision to commit units intended for the retaliatory offensive to the battle to

157

prevent a Soviet breakthrough. If such a scenario developed, NATO may militarily be unable to undertake the planned offensive.

On the negative side, a heavier first echelon attack may present an extremely tempting target for NATO tactical nuclear weapons, particularly if the FEBA were about to crumble. Even if the FEBA held, Soviet casualty rates probably would be higher simply because the heavier first echelon would present such a "target rich" environment.

If, because of these or other considerations, the USSR were to opt against a heavier first echelon attack, it may choose to increase its emphasis on the operational maneuver group (OMG), that is, a self-contained unit of division or larger size that is highly mobile, heavy in armor, and capable of deep penetration.[14] OMGs could be used for independent actions, first or second echelon support, or counteroffensive operations. The mobility, firepower, and independence of OMGs make them particularly attractive from a planner's perspective; whether they would be either as mobile or as independent under combat conditions has not been tested. Nevertheless, Soviet planners may find them an attractive response to a NATO retaliatory offensive strategy because of their perceived flexibility. OMGs could be used either for deep penetration to destroy primary targets more rapidly and from the Soviet perspective bring the war to a satisfactory conclusion before NATO could employ its offensive strategy, or they could be used against the assembly areas from which NATO planned to launch its offensive.

Static Defenses. By its very nature, a NATO retaliatory offensive strategy must commit units into territory that the Warsaw Pact has had time to prepare to its own liking. This preparation could include defensive mining, tank traps and other hazards, and demolitions. Indeed, if as some Western analysts have maintained, one of the Soviets' greatest fears in an offensive is remote mining,[15] it is likely that the USSR and its Warsaw Pact allies would turn to this and other forms of static defense to frustrate a NATO retaliatory offensive strategy.

The key to a static defense obviously lies in identifying an enemy's line of advance. Given the planned mobility of modern warfare, this will prove a difficult task for Soviet planners, even as it is for NATO planners. But certain lines of advance are more attractive than others, and these would be the areas most probably prepared with static defenses, if only to channel advancing NATO forces into fire sacks targeted by Warsaw Pact elements reserved for defense. While no Soviet planner is likely to concentrate solely on static defense, it could prove a useful adjunct to other Soviet military responses against a NATO retaliatory offensive strategy.

Chemical and Nuclear Options. Before NATO could launch a retaliatory offensive, it would have to concentrate its forces designated to participate in the offensive. As already discussed, such an assembly area may prove a tempting target for OMGs. Inevitably, it may also be a tempting target for Soviet chemical or nuclear weapon use simply because of the concentration of forces.

From the Soviet perspective, chemical and nuclear first use have both advantages and disadvantages. While chemical or nuclear weapons could undoubtedly degrade the NATO offensive, they probably also would eliminate any unwillingness within NATO to use nuclear weapons against Warsaw Pact forces. Although it is doubtful if politically opportunistic peacetime promises of "no first use" would be meaningful during war, particularly if the "first user" could obscure which side first resorted to nuclear weapons and the "first user" were convinced he could obtain a military advantage, one cannot dismiss the possibility that Soviet planners may be more hesitant to cross the nuclear threshold because of this promise. However, offsetting this hesitancy is the fact that an early first use of nuclear weapons against a building NATO offensive concentration would take place on NATO, not Warsaw Pact, soil. Somewhat ironically, the USSR may be pushed into early first use if NATO successfully halts the original Warsaw Pact offensive or launches its own retaliatory offensive plans. In a classic case of the shoe being on the other foot, Soviet planners, even as NATO planners, may in this situation believe they have no choice other than to use nuclear weapons or watch enemy forces break through. Any Soviet use of chemical weapons against building NATO offensive concentrations would follow similar lines of logic. However, because of lower levels of collateral damage, pressures to employ chemical weapons early against the building concentration would be less.[16] Most chemical weapons would cause less damage to the surrounding territory than nuclear weapons, and the USSR may therefore be more willing to delay their use until NATO forces crossed into Warsaw Pact territory.

RETALIATORY OFFENSIVE STRATEGIES
AND DETERRENCE

We have argued that the Soviet Union has a variety of options to choose for its response to a NATO retaliatory offensive strategy. Some appear better than others; each could be undertaken in conjunction with others. Indeed, if past Soviet behavior in response to NATO initiatives is any indication, the Soviet Union would respond to NATO's hypothetical new strategy with a broad front of actions and threatened actions.

What impact, then, would a new NATO strategy and Soviet responses to that strategy have on deterrence? At the very least, a retaliatory offensive strategy must create several new levels of uncertainty in the minds of Soviet planners. Would NATO really move east? Could Warsaw Pact forces stop a NATO conventional retaliatory offensive? How would non-Soviet Warsaw Pact forces and peoples react if NATO units moved into their countries? Could Soviet forces defend their logistic lines? To the extent that uncertainty adds to deterrence, deterrence would be enhanced.

But how much would Soviet uncertainty be increased? Any change in NATO strategy would be subject to considerable debate in the West, and the possibility exists that even if a conventional retaliatory offensive strategy were adopted, its implementation during time of war would remain hostage to the vagaries of alliance political decisionmaking even as nuclear release is. If Soviet decisionmakers concluded that NATO would probably not implement its new strategy because of such considerations, deterrence would not be enhanced.

Could Warsaw Pact forces stop NATO's offensive? If Soviet planners concluded that they could, then the retaliatory offensive strategy would contribute little to deterrence unless Soviet decisionmakers estimated that the cost to the USSR of stopping NATO's advance was too great to bear. Thus from NATO's perspective, it is imperative that adoption of a conventional retaliatory offensive strategy create in the minds of Soviet decisionmakers either the perception that NATO's offensive cannot be stopped or that the cost of stopping it would be greater than the benefits expected from initiating conflict. And to create either perception, particularly in light of potential Soviet military doctrinal and tactical responses, a significant enhancement of

NATO's conventional military capabilities would be required. If Soviet decisionmakers concluded that NATO's forces were insufficient to achieve either end, deterrence would not be enhanced.

How would non-Soviet Warsaw Pact forces and peoples react if NATO units moved east? East European mistrust and even hatred of their Soviet overlords cannot be equated with support for and adherence to values and concepts prevalent in the West. The Soviets may not be wanted in Warsaw or Prague, but it is far from clear that Germans or Americans are, either. Germany in particular has not had a pleasant history with its eastern neighbors, and as far as the United States is concerned, it is worth remembering that the "Successor Generation" problem may not be limited to Western Europe. East European populations may still detest the Russians as much as they ever did, but they have been fed a steady diet of anti-Americanism and anticapitalism for over a third of a century. German and American tanks moving toward Warsaw, Prague, or Budapest may draw the same response from the local citizenry as Russian tanks would. This clearly would not add to Soviet uncertainties about initiating a Warsaw Pact offensive against Western Europe.

Even if conflict were never to break out, it is not clear that adoption of a new NATO military strategy would lead to an East European effort to maximize political distance between themselves and Moscow. Such an assumption requires one to believe that East European leaders would both interpret the new NATO strategy to imply that NATO forces would aid them if Moscow invoked the Brezhnev Doctrine, and to interpret the new strategy as presenting no threat to them. Neither interpretation is likely.

From the Soviet perspective, then, the most effective counter to a NATO retaliatory offensive strategy may well be a widespread political-diplomatic offensive throughout Western and Eastern Europe and the United States that stresses the aggressive nature of NATO, the new dangers of war, and the disparity of US and European interests. This political-diplomatic offensive would be followed shortly thereafter by renegotiation of status of forces agreements, restationing of Soviet troops from the USSR's European Military Districts to Eastern Europe itself, and some preparation of static defenses. Such a combined Soviet response, if handled skillfully, could raise doubts in Western Europe about

NATO's purpose and American intent, increase East European perceptions of a "NATO threat," and militarily allow the Warsaw Pact to counter NATO's retaliatory offensive strategy without any significant changes in the USSR's military strategy. Whether such a chain of events, when seen from NATO's perspective, would enhance deterrence in Europe is doubtful.

Outside Europe, a NATO retaliatory offensive strategy contributes nothing to deterrence of Soviet actions. It is simply not credible politically. Would Western leaders sanction a NATO retaliatory strike into Eastern Europe in response to a Soviet invasion of Iran or Saudi Arabia, or in response to Soviet interventions elsewhere in Africa or Asia? Probably not. A NATO attack on Eastern Europe in support of China in the event of a Sino-Soviet war is similarly not plausible. The level of conflict escalation in all these cases is simply too great. This is not to say that such escalation may not be desirable. However, the political environment that exists in the United States and Western Europe means that no decisionmaker—and therefore no NATO commander—could credibly make that linkage and survive politically.

In another area, claims that a NATO retaliatory offensive strategy would lead, in Soviet eyes, to increased pressures to attack during European crisis situations appear overstated. Assuming that NATO states of readiness remain tied to but lag behind Warsaw Pact states of readiness, no post-strategy-change decrease in crisis stability would exist. If, however, a NATO retaliatory offensive strategy required that NATO divisions be placed on higher readiness levels at earlier stages of the crisis, some decrease in crisis stability would occur. How much stability would decrease would be a function of the Kremlin's perception of NATO's ability to launch an initiatory offensive, not a retaliatory offensive.

All things considered, then, given probable Soviet perceptions of and likely Soviet responses to a NATO retaliatory offensive strategy, it is far from certain that such a strategy would contribute meaningfully to deterrence in Europe. Neither would such a strategy have deterrent effect beyond Europe. Nevertheless, if only because NATO's current strategy has serious shortcomings, a conventional retaliatory offensive strategy does warrant further study. Additional study must center first on two significant political and military conditions. If these conditions are not convincingly met, then thoughts of a NATO retaliatory offensive strategy must be totally and completely discarded.

First, NATO must be certain that the mere discussion of a change to its military strategy will not become a disruptive force within the Alliance. Western body politics are much less prone to react to changes in strategy than to changes in weaponry. Nevertheless, such a significant change as contemplated by a move from flexible response to a conventional retaliatory offense may be expected to engender great criticism and loud debate that the Kremlin would seek to bend to its own ends. Thus, if a retaliatory offensive strategy becomes a serious strategy alternative, NATO and its member governments must couch the debate in careful terms and define it clearly as a response to a Soviet first offensive. To do otherwise would be to undermine further the consensus on which NATO is based and to weaken more the deterrent utility that NATO currently has. Neither would be in Western interests.

Second, a retaliatory offensive strategy would require significant additional conventional forces if it were to be effective even in the absence of Soviet military responses. In the presence of Soviet responses, the need for even more forces is apparent. Without requisite political will and economic commitment to provide those forces, contemplation of a NATO retaliatory offensive strategy is meaningless, and a strategy without the capabilities to carry it out is both dangerous and irresponsible.

ENDNOTES

1. Western estimates of the growth rate of Soviet military expenditures differ widely. One CIA estimate places the 1970-80 annual growth rate at three percent, while one of William Lee's estimates is as high as ten percent. Other American, British, Chinese, and French estimates fall between these extremes. See International Institute for Strategic Studies, *The Military Balance 1982-1983* (London: International Institute for Strategic Studies, 1982), pp. 12-13.

2. For some recent examples, see *Pravda,* August 11, 1980, and November 17, 1982; *Izvestiya,* December 16, 1982; and B. Halosha, "How an Aggressive Bloc Works," *World Marxist Review,* vol. 25, no. 3 (March 1982), pp. 67-69.

3. See *Pravda,* August 3, 1982 and October 24, 1982; *Izvestiya,* July 5, 1980; and V. Yulin and V. Gurevich, "East-West Trade and the Helsinki Principles," *Ekonomicheskaya gazeta* (November 1980), p. 21, for just a few of the many possible examples.

4. "Letter by Brezhnev Warns NATO Lands on the Neutron Bomb," *New York Times,* January 24, 1978, p. 3, and "NATO Nations to Reply to Soviet," *New York Times,* January 28, 1978, p. 48.

5. See, for example, *Pravda,* February 7, 1983, February 16, 1983, March 20, 1983; and *Izvestiya,* March 1, 1983, March 2, 1983, and March 18, 1983.

6. For a discussion of some of the causes of the "Successor Generation" situation, see David A. Andelman, "Struggle Over Western Europe," *Foreign Policy,* no. 49 (Winter 1982-83), pp. 37-51. See, also, Arrigo Levi, "Western Values and the Successor Generation," *NATO Review,* vol. 30, no. 2, 1982, pp. 2-7.

7. For discussions of West European and NATO attitudes concerning European and NATO force deployments beyond Europe, see Drew Middleton, "Allied Attitude on a Gulf War Troubling U.S.," *New York Times,* February 17, 1980, p. 11; "NATO Chief Against Wider Role," *New York Times,* June 18, 1980, p. 2; John Vinocur, "Bonn Expects Pressure from Allies to Send Warships to Persian Gulf," *New York Times,* October 13, 1980, p. 16, and Drew Middleton, "Massed Allied Warships: Warning to Iran on Strait," *New York Times,* October 17, 1980, p. 14.

8. See, *Pravda,* August 28, 1980, September 4, 1980, September 20, 1980, and December 18, 1980. Many other examples exist.

9. See, for example, "Growing Danger of New World War," *Peking Review,* vol. 19, no. 2 (January 9, 1976), pp. 17-18; "Soviet Social-Imperialism—Most Dangerous Source of War," *Peking Review,* vol. 19, no. 5 (January 30, 1976), pp. 9-13; "Intensified Superpower Contention for Hegemony in Western Europe," *Peking Review,* vol. 19, no. 11 (March 12, 1976), p. 15; and "Western Public Opinion Senses Danger in Nurturing Tiger," *Peking Review,* vol. 19, no. 15 (April 9, 1976), pp. 15-17.

10. See David Binder, "Ceausescu Resisting Rise in Arms Outlay," *New York Times,* November 28, 1978, p. 15; and David A. Andelman, "Poland to Freeze Arms Budget Despite Soviet Request," *New York Times,* January 9, 1979, p. 3, for details of Rumanian and Polish opposition to increased defense spending. By contrast, East Germany has forged ahead with higher expenditures. See "East Germans Expand Military Expenditures as 'Counter' to NATO," *New York Times,* December 18, 1979, p. 5; and "East Germany Announces Increase of 8.4% in '81 Military Spending," *New York Times,* December 18, 1980, p. 10.

11. The so-called Brezhnev Doctrine was proclaimed by the USSR to justify its 1968 invasion of Czechoslovakia and declares that it is the right and duty of socialist states to defend socialism in other socialist states. Although it has not been formally invoked to justify military intervention since Czechoslovakia, the Brezhnev Doctrine clearly seeks to legitimize Soviet intervention, if needed, throughout the socialist world.

12. For discussions of Soviet echeloning practices, see US Department of the Army, *Soviet Army Operations* (Washington: US Army, 1978), especially Chapter 3; P. H. Vigor, "Soviet Echeloning," *Military Review,* vol. 62, no. 8 (August 1982), pp. 69-74; Trevor N. Dupuy, "The Soviet Second Echelon: Is This a Red Herring?" *Armed Forces Journal International,* vol. 119, no. 12 (August 1982), pp. 60-64; and Mark Stewart, "Second Echelon Attack: Is the Debate Joined?" *Armed Forces Journal International,* vol. 120, no. 1 (September 1982), pp. 105-108 and *passim*.

13. US Army, FM 100-5, *Operations* (August 20,1982), *passim,* and Bernard W. Rogers, "Sword and Shield: ACE Attack of Warsaw Pact Follow-On Forces," *NATO's Sixteen Nations,* vol. 128, no. 1 (February-March 1983), pp. 16-26. See, also, Chapter 3 and Chapter 4 in this volume for more information on AirLand Battle and Follow-On Force Attack.

14. See C. N. Donnelly, "The Soviet Operational Maneuver Group: A New Challenge for NATO," *International Defense Review,* vol. 15, no. 9 (September 1982), pp. 1177-1186; and David M. Glantz, "Soviet Operational Formation for Battle: A Perspective," *Military Review,* vol. 63, no. 2 (February 1983), pp. 2-12.

15. See C. N. Donnelly, "The Soviet Operational Maneuver Group," p. 1186.

16. For a discussion of comparative Soviet and American chemical warfare capabilities, see J. P. Perry Robinson, "Chemical Weapons and Europe," *Survival,* vol. 24, no. 1 (January-February 1982), pp. 9-18.

CHAPTER 8

DOES THE UNITED STATES NEED A NUCLEAR WARFIGHTING DOCTRINE AND STRATEGY?

by

Keith B. Payne

The evolution of American strategic doctrine since the early 1970s is said to have been a transition away from deterrence, towards a so-called "warfighting" doctrine. Discussion of this transition was initiated by the "Schlesinger Doctrine" and National Security Decision Memorandum (NSDM) 242 in 1974,[1] and continued under each successive administration: best illustrated by Presidential Directive (PD) 59 during the Carter Administration,[2] and National Security Decision Directive (NSDD) 13 under the Reagan Administration.[3] US doctrine during this transition has emphasized the capability for limited strategic options, countermilitary and counterpolitical control targeting, postattack continuity of government, and the potential for waging a prolonged nuclear conflict, *inter alia*. In effect, this so-called "warfighting" transition has focused primarily upon revising US nuclear targeting policy, and emphasizing "counterforce" targeting even more than had been the case previously.

These new emphases in US strategic doctrine are in direct contrast to the so-called mutual assured destruction (MAD) approach to deterrence which envisages mutual societal vulnerability as the basis for deterrence, and counterforce targeting as destabilizing. Some have gone so far as to label those supporting this transition as "warfighters," and those who endorse a redirection of US strategic thinking back towards MAD as "stable balancers."[4]

The use of such labels as "warfighting doctrine," "MAD," "warfighters," and "stable balancers" generates more smoke than understanding of the American strategic debate. Indeed these names are more confusing than descriptive; so-called warfighters are no more interested in fighting a nuclear war than are proponents of MAD interested in ensuring mutual destruction. All schools of strategic thought in the United States consider the deterrence of war to be the primary objective of nuclear policy.

To identify any particular approach to strategic nuclear deterrence as "warfighting" is a mistake. Strategic deterrence must envisage the possibility that nuclear weapons may be used—a nuclear deterrent is hollow without an implicit or explicit threat of waging nuclear war under some conditions. Whether that threat is based upon MAD or counterforce targeting, nuclear deterrence is an inherently "warfighting" concept. Indeed, one of the primary deficiencies of the Bishops' Pastoral Letter is the recognized need for nuclear deterrence, yet the absolute reluctance to acknowledge the associated need for the threat of nuclear use.[5] The Bishops' movement appears to favor the stability of nuclear deterrence, while opposing the nuclear threat which underlies that stability.

In short, all approaches to nuclear deterrence are predicated upon the threat of nuclear weapons use under certain conditions, i.e., "warfighting." Consequently, the answer to the question posed above, "does the United States need a warfighting strategy and doctrine," must be yes, if the United States is interested in the stabilizing effects of nuclear deterrence.

However, if one accepts common parlance, equating the development of tailored counterforce planning (e.g., NSDM-242 and PD-59) with "warfighting," and MAD as an alternative to warfighting, then the issue of which direction should guide US policy becomes more complex. This chapter will examine the different approaches to deterrence and assess which is most appropriate as the basis of American nuclear doctrine and strategy.

AMERICAN STRATEGIC THOUGHT: THREE APPROACHES TO STRATEGIC NUCLEAR DOCTRINE

There are three distinct schools of American thought regarding nuclear doctrine. Such a generalization of American strategic thought, by definition, does some damage to nuance; and additional categories and subcategories could be identified.[6] However, these three encompass the major tenets of policy relevant thought: MAD; a "warfighting deterrent strategy"; and a balanced offensive-defensive approach to deterrence. Each of these can be reviewed briefly.

MAD. Mutually assured destruction is predicated upon a relatively simple line of reasoning: stability results when both sides possess strategic nuclear retaliatory forces and vulnerable homelands. It is argued that neither would choose to engage in nuclear first-use or extremely provocative actions (that might escalate to nuclear war) because no political-military objective could be worth the cost of homeland destruction.

There are certain problems with the MAD approach to deterrence.[7] For example, it is based entirely upon the assumption that the Soviet leadership will behave rationally. That is, the Soviet leadership will rationally assess the risks and potential costs of nuclear use or highly provocative behavior (such as an invasion of NATO Europe) and conclude that the costs would outweigh the potential benefits of aggression. Unfortunately, history does not lack for national leaders pursuing apparently irrational modes of behavior; MAD makes no provision for such a contingency.

Similarly, MAD makes no provision for the possibility that the Soviet leadership could take extreme military measures in response to an extreme situation, with full appreciation of the risks involved. Highly provocative actions, or even nuclear use could be viewed as the least miserable option during an intense national crisis. Professor Samuel Huntington has noted that the Soviet Union may well be confronting such extreme internal or external crises during the decade.[8]

The uncertainties involved in any nuclear escalation, and the risks entailed by any nuclear use are suggested as factors that would *always* deter Soviet nuclear use. However, under extreme duress the Soviets may see those uncertainties, the "fog of war," as holding forth the slim hope for escaping their crisis through a military initiative, and therefore be willing to accept the associated risks.

As the Soviet Union responds to the disintegration of its domestic and extended empire, it could react irrationally or with extreme military force.[9] MAD makes no provision for either occurrence; indeed, because it endorses a condition of absolute societal vulnerability, it virtually ensures maximum destruction in the event deterrence fails.

There is some recent precedent for this type of decisionmaking that is beyond deterrence: the Japanese appear to have entered World War II with a view that war with the United States was very risky, and yet the least miserable option. As Louis Morton observes in the US Army's history of World War II decisionmaking:

> In the final analysis, the Japanese decision for war was the result of the conviction, supported by the economic measures imposed by the U.S. and America's policy in China, that the U.S. was determined to reduce Japan to a position of secondary importance. The nation, Tojo and his supporters felt, was doomed, if it did not meet the challenge. In their view, Japan had no alternative but to go to war while she still had the power to do so. She might lose, but defeat was better than humiliation and submission. 'Japan entered the war,' wrote a prince of the Imperial Family, 'with a tragic determination and in desperate self-abandonment.' If it lost, 'there will be nothing to regret because she is doomed to collapse even without war.'[10]

The point here is not that the Soviet position during the 1980s will be completely analogous to that of pre-World War II Japan—although there may well be similarities. The point is that in the context of acute national crises Soviet decisionmaking may proceed along a course that is not adequately addressed by the logic of MAD.

In addition, MAD is not appropriate for the broad range of American deterrence responsibilities. For example, a primary function of US strategic nuclear forces is to extend deterrence coverage to America's European allies. According to MAD proponents, the possibility that a major war in Europe could escalate uncontrollably to strategic nuclear war is threat enough to deter the Soviet Union from even conventional war in Europe.[11]

One of the traditional rationales for not matching the Soviets in conventional military capabilities is that nuclear deterrence can compensate for deficiencies in nonnuclear forces.[12]

However, if the United States is vulnerable to Soviet nuclear retaliation of any scale, the credibility of an American threat to engage in strategic nuclear war on behalf of distant allies should not be high. "Extended deterrence" in the context of absolute US societal vulnerability is, in effect, a threat to commit national suicide to preserve distant interests that are extremely important, but are not immediate survival interests. A country may be able to wield a suicidal deterrent threat credibly in support of immediate and direct survival interests, but that credibility should be lacking in relationship to nonsurvival interests. It is difficult to believe that any American President would purposefully choose to escalate even an unfolding defeat in Europe—a defeat which would entail dangerous implications indeed for the United States—into a central strategic war which would threaten American survival. Such a suicidal threat is incredible for such a deterrent role, and is therefore inappropriate as the basis for US extended deterrence responsibilities.

In short there are three clear inadequacies of the MAD approach to deterrence. It does not address the potential for irrationality on the part of the Soviet Union; it does not address the distinct possibility that the Soviet leadership could, during an acute crisis, consider the use of extreme military force to be the least miserable option— in full recognition of the risks of escalation to nuclear war involved (MAD endorses the condition of vulnerability that would virtually ensure the maximum level of catastrophy if deterrence fails for either of these reasons); and it is inappropriate to support the broad range of US deterrence responsibilities.

Warfighting/Counterforce. The second school of strategic thought is the so-called "warfighting" approach to deterrence. The concepts of counterforce targeting and limited nuclear options, the two primary emphases of the "warfighting" approach, have been discussed for several decades.[13] These concepts certainly were not originated by the Nixon, Ford, Carter, or Reagan Administrations, although they appear to have been endorsed officially during the tenure of Dr. James Schlesinger as Secretary of Defense, and reinforced with somewhat different degrees of emphasis by each succeeding administration. The warfighting label has been applied in particular to several official strategic initiatives spanning almost a decade: NSDM-242 (1974), PD-59 (1980), and NSDD-13 (1981).

Proponents of a "warfighting" approach to deterrence suggest several reasons why it is superior to MAD as the basis of US policy. First, it was at least suggested by Secretary of Defense Harold Brown that the counterforce orientation of PD-59 would provide a more fearsome deterrent, from the Soviet perspective, than an indiscriminant threat to societal assets; it was said to threaten what the Soviets "really value." As Brown noted regarding PD-59:

> The biggest difference I would say, that PD-59 introduces is a specific recognition that our strategy has to be aimed at what the Soviets think is important to them, not just what we think would be important to us in their view.[14]

Second, it was and is suggested that the counterforce and counterpolitical control targeting of a "warfighting" strategy could degrade Soviet military, and particularly strategic nuclear capabilities in the event deterrence fails. Consequently, a "warfighting" strategy has been seen as a return to the traditional military mission of attriting the opponent's military capabilities, and thereby perhaps limiting wartime damage to the United States and its allies.

Finally, the concept of limited nuclear options (LNOs) that was a principal tenet of the Schlesinger doctrine and NSDM-242 represented, in part, an attempt to enhance the credibility of the American extended deterrent.[15] It was argued, rightly, that if the United States could retaliate strategically only in a massive fashion to a Soviet attack on NATO Europe, then the United States might be deterred from responding strategically given the Soviet capability to retaliate against the US homeland in a similar fashion. In such a context, the US strategic extended deterrence would be viewed as incredible.

Limited nuclear options were seen as a means of enhancing the credibility of the US extended strategic deterrent. The argument was that a limited nuclear threat would provide a more credible extended deterrent because the Soviet Union would be unlikely to retaliate massively against the American homeland in response to a limited US strike. A limited American nuclear strike would leave the Soviet Union with many assets left at risk and hence be a major disincentive to retaliate massively against the United States.

171

Limited nuclear options were viewed as a means of coercing the Soviet Union to engage in targeting restraints and limitations even in the context of central war—extending deterrence into the intrawar phase of central war. As such, LNOs were considered a means of circumventing, through a change in strategy, the suicidal character of the US extended deterrent. Presumably, an extended deterrent of a less suicidal nature is a more credible deterrent.

There are three primary inadequacies of the "warfighting" strategy and doctrine described above. Surprising as it may seem, such a "warfighting" approach to deterrence shares several inadequacies in common with MAD. For example, as is the case with MAD, a "warfighting" approach to deterrence does not address the potential for irrationality on the part of Soviet decisionmakers. It is predicated upon the same assumption of an opponent making rational cost-benefit calculations. Indeed, warfighting assumes a somewhat more sophisticated analysis by the Soviet leadership as it is expected to take into rational consideration the logical implications of US limited nuclear options. (Unfortunately, some evidence suggests that US LNOs would not have the hoped-for restraining effect on Soviet targeting in a central war.[16])

Similarly, the "warfighting" approach to deterrence makes no provision for the possibility that the Soviet leadership could, at some time, consider war with the West to be the least miserable option. In terms of providing security for the United States, "warfighting" offers little useful guidance in the event that the Soviets are beyond deterrence. In short, neither MAD nor the so-called "warfighting" approach to deterrence make provision for the failure of strategic deterrence for any reason.

Second, although it often is asserted that the Soviet leadership would consider a threat to its military and political power to be more deterring than an indiscriminate threat to Soviet industrial and societal assets, little direct evidence supports such a hypothesis. Although it virtually has become accepted wisdom, little evidence is offered to support the notion that a countermilitary and counterpolitical threat would be more deterring—other than the typical reference to the instrumental value Soviet leaderships have placed on the bulk of the Soviet population.[17] However, noting that Soviet leaders have behaved in a despicable fashion toward large numbers of Soviet citizens is a far cry from offering evidence that a threat to destroy a very large percentage of Soviet industrial, societal, and soft military assets is less deterring from the Soviet perspective than a more discrete threat against Soviet military forces, and military and political control assets. This is not to say that the latter threat would not be more deterring (indeed, given the general understanding we have of the Soviet Union and its leaders it seems correct, intuitively, that a countermilitary threat would be more deterring). However, the point here is simply to note that such a notion has become conventional wisdom without reference to directly relevant supporting evidence.

Of course, given the nature of the subject, marshaling evidence to "prove" that one form of threat is more deterring than another is an extremely tall order. The likely perceptions and values of a foreign leadership during the highly complex and unusual circumstances of a nuclear crisis are not phenomena that can be subjected easily to any scientific method of proof.

Third, the notion that LNOs significantly enhance the credibility of the US extended deterrent also is open to considerable question. It would appear unlikely that either the apparent counterforce emphasis of PD-59 and NSDD-13, or the apparent LNO emphasis of NSDM-242 can enhance the credibility of the US extended deterrent. The lack of credibility for extended deterrence is not a function of US targeting plans or options per se. Rather, the lack of credible strategic deterrent coverage for distant allies and interests is a function of the absolute vulnerability of the American homeland to Soviet nuclear attack.[18] The important question is, and has been: how can the United States gain deterrence leverage for distant interests when the result of implementing the US strategic deterrent threat would likely be national destruction? Adjusting offensive targeting plans (even in very sophisticated ways a la PD-59) cannot provide an answer to that question. As Herman Kahn noted over twenty years ago:

> But so far as the Soviets are concerned, the probability of a [nuclear escalation] by us is small, particularly because we have made negligible preparations to ward off, survive, and recover from even a 'small' Soviet retaliatory strike. No matter how menacing they look it will be irrational to attack and thus insure a Soviet retaliation unless we have made preparation to counter this retaliation.[19]

Such an observation is even more relevant now than it was when first written, given the vast increase in Soviet strategic nuclear capabilities. Why would any American President escalate a war taking place outside the American homeland to a war involving nuclear attacks on the homeland, with the potential for scores of millions of American casualties (even if that escalation began with a "limited" use of nuclear weapons)? As Donald Brennan often noted, in the event of war with the Soviet Union the American President would be much more interested in saving American lives than inflicting Russian casualties;[20] such a reasonable perspective would argue strongly against US strategic nuclear escalation, whether limited or not. How then can LNOs enhance the credibility of an American extended deterrent, a deterrent which is based upon the threat of nuclear escalation?

The critical problem of American vulnerability, and hence the incredibility of the extended deterrent, cannot be ameliorated by adjusting targeting plans to emphasize strongly counterforce targeting, or LNOs, or both. Consequently the "warfighting" emphasis upon counterforce targeting and LNOs cannot solve the problem of an incredible extended deterrent.

The so-called "warfighting" school of thought outlined above cannot enhance the credibility of extended deterrence unless it envisages effective damage-limitation for the American homeland to counter the destruction that could result from an emphasis on counterforce targeting and LNOs. Yet, it is improbable that offensive targeting alone, even if it is heavily counterforce, could significantly limit damage to the American homeland in the event of central war. Unfortunately, such an offensive "warfighting" strategy could be even more dangerous than MAD. Limited nuclear options and counterforce targeting could make nuclear escalation appear somewhat less risky to an American President, yet it would not be likely to ameliorate the consequences of engaging in such escalation. As BG Albion W. Knight, Jr. (USA Ret) observed regarding the "warfighting"orientation of PD-59: ". . . U.S. citizens continue to be left hostage to a Soviet strategic attack under the new doctrine."[21]

In brief, what has come to be known in the popular jargon as a "warfighting" strategy shares the same inadequacies as MAD: it cannot address the potential for irrational behavior, or the possibility that extreme military force could be considered the least miserable option during an intense crisis; nor is it appropriate as the basis for meeting American extended deterrence responsibilities. In addition, little convincing evidence can be marshaled to support the notion that a robust counterforce threat would be more deterring to the Soviet leadership than a MAD-oriented threat.

A BALANCED OFFENSIVE/DEFENSIVE DETERRENT

The above discussion illustrates why an adequate approach to deterrence must consider strategic defensive forces and damage-limitation. Adjusting US offensive targeting plans and the so-called warfighting strategy address only half of the problem. This observation leads to the third school of thought—a balanced offensive/defensive deterrent.[22]

This balanced approach to deterrence envisages a combination of the offensive capabilities of a "warfighting" strategy, and active and passive strategic defenses intended to minimize the vulnerability of the American homeland. Such active and passive strategic defenses might include: a boost phase, space-based high energy laser (HEL) system; midcourse (and high endoatmospheric) land or space-based nonnuclear kill (NNK) interceptors; endoatmospheric and terminal phase land-based NNK interceptors; modernized air defenses; and increased planning for population and industrial passive defense. Of course this entire comprehensive multilayered defensive system will not be available in the near future. Rather, such a defensive transition could require several decades or longer to accomplish.

A balanced deterrent and force posture should address those inadequacies of MAD and "warfighting." For example, a balanced deterrent should enhance the credibility of the US extended deterrent. An America less vulnerable to nuclear attack should be perceived by friend and foe as an America more willing to stand by its allies and treaty commitments, even under the most stressful conditions. Damage-limitation would, in effect, lower the risk, or at least the perception of risk, the United States would have to accept as a result of extending strategic deterrent coverage to distant interests and allies.

A balanced deterrent should also greatly increase Soviet first-strike uncertainties, thereby strengthening deterrence stability. Effective active defense for US retaliatory forces would reduce the Soviet capability to change the "correlation of forces" during the first phase of a central war—a critical requirement in the Soviet so-called theory of victory.[23] Reducing Soviet first-strike confidence should considerably improve crisis stability.

Finally, a balanced deterrent represents the only school of thought that makes provision for the possible failure of deterrence. To emphasize offensive planning alone, in accordance with MAD or "warfighting" strategy, is to place virtually all of the potential for limiting damage to the American homeland on the continued functioning of deterrence. However, as previously discussed, deterrence could fail as a result of the opponent's irrationality, or a perception that war is the least miserable option. The probability of the latter occurring may increase as the Soviet Union confronts increasing pressures for devolution of its extended empire during the 1980s.[24] Only a balanced deterrent, with its relatively greater emphasis on damage-limitation, addresses this possibility by making provisions for American security in the event deterrence fails.

There are, to be sure, some potential problems involved with a balanced deterrent. For example, effective programs for damage-limitation must be comprehensive. A commitment to Ballistic Missile Defense (BMD), such as that suggested by President Reagan on March 23, 1983[25] would make little sense unless it were complemented by modernized air defenses, and increased civil defense planning. Deploying any single element of a defensive package would provide the Soviet Union with incentives to channel its offensive efforts around that individual defensive system. For example, a massive and very effective American BMD system would be likely to redirect the Soviet offensive emphasis toward bombers and cruise missiles. Consequently, modernized air defenses are a logical complement to BMD.

Interestingly, past critics of US strategic air defense suggested that because the United States could not defend against ballistic missiles, there was little sense in maintaining American air defenses. More recently the argument has been reversed. It is now suggested that BMD would be of little utility because the Soviets would bypass it with cruise missiles and bombers. In fact, a commitment to damage-limitation would necessitate defenses against air-breathing and ballistic threats.

Such a comprehensive defensive program would encounter significant obstacles. It would require a significant commitment of always scarce defense dollars; and it would, eventually, require revision of, or withdrawal from the ABM Treaty—a treaty some Americans consider "the most successful arms limitation agreement ever . . . [and] even attempts to amend the Treaty in a supposedly minor way can be treacherous."[26]

A comprehensive defensive transition involving space-based beam weapons may also run into technical difficulties, even if funded adequately.[27] Nevertheless, land-based endo- and exoatmospheric NNK BMD would be technically less risky, and less costly than space-based defenses. These nearer-term BMD systems, while not holding the promise for a comprehensive protection of American urban and industrial assets, could contribute significantly to the survivability of US strategic forces, and National Command Authorities (NCA) and Command, Control, Communications and Intelligence (C^3I) assets.[28]

Some of the critical strategic retaliatory assets suitable for initial BMD coverage include:[29]

- Alternate National Military Command Center (ANMCC).
- Minuteman ICBM launchers.
- Minuteman Launch Control Centers (LCCs).
- Emergency Rocket Communications Systems (ERCS).
- National Emergency Airborne Command Post (NEACP).
- NEACP and Crown Helicopter rendezvous sites.
- Post-Attack Command & Control System, Air Launched Command Center (PACCS ALCC).
- Early Warning Satellite, Mobile Ground Stations (MGS).
- National Command Authorities and Joint Chiefs of Staff, National Military Command Center (NCA/JCS NMCC).

Enhancing the survivability of US retaliatory capabilities through deployment of near-term, land-based BMD systems as the first phase of a defensive transition should help ensure stability during a transition to a comprehensive defense. If a complete transition is cut short for technical or political reasons, such a first phase would still be valuable in and of itself.

It should be assumed that if the United States seriously pursues strategic defensive forces, the Soviet Union will increase its already substantial effort in the area of active and passive defense,[30] including the deployment of BMD. Potential crisis instability could arise from a reduced American capability to maintain a deterrent threat against the Soviet Union if it complements existing counterforce offensive capabilities, and existing air and civil defense preparations with nationwide BMD.

However, if the United States provides active defense for its strategic retaliatory forces (and selected NCA and C^3I facilities), it will increase the US retaliatory potential that would survive Soviet counterforce attack, and thereby could compensate partially or wholly for Soviet BMD deployment and the resultant reduction in America's capability to threaten Soviet targets. BMD coverage for selected US strategic forces, and NCA and C^3I assets should help sustain deterrence stability even assuming advanced Soviet BMD deployment. In short, US deployment of land-based BMD for selected strategic retaliatory assets and particularly NCA and C^3I facilities would be supportive of a longer-term comprehensive defense by ensuring the US deterrent, and thus stability, during the initial phases of a transition period; in the longer term it would provide two important components of a multilayered comprehensive defensive architecture. Given the existing vulnerabilities of US NCA and C^3I assets, BMD deployment for the purpose of defending these critical points would be extremely beneficial for deterrence even if the United States does not complete the transition to a comprehensive strategic defense.[31]

PUNITIVE VERSUS DENIAL DETERRENCE

A balanced deterrent, including comprehensive homeland defense, would involve some important differences from the current "punitive" US-Soviet deterrence relationship. A balanced deterrent would constitute a denial deterrent rather than a punitive deterrent. The difference between the two is significant.

A punitive deterrent, whether MAD or "warfighting" oriented, gains its deterring effect by threatening to punish the opponent if it takes an extreme course of action. That threat is levied against what is thought to be the opponent's highly valued assets. The intent of a punitive deterrent is to ensure that the opponent will consider the cost of US retaliation to outweigh the potential benefit of the unwanted course of action. Limiting damage to the United States is not part of these punitive approaches to deterrence; ensuring that costs outweigh benefits involves threatening what the opponent values, not defending what Americans value. The basic "punitive" line of reasoning is: when the highest Soviet values are at risk, then no political-military objective could be worth the cost that would be inflicted by US retaliatory action. Hence, given the potential for such punishing US retaliation, the Soviet Union will be deterred from highly provocative actions.

A denial deterrent involves a different basis for its deterrent effect. The objective is still to ensure that the cost of provocative behavior outweighs potential gains, yet the cost is not based entirely upon threatening to punish the Soviets by launching a retaliatory strike against that which Soviet leaders value most. Rather, the costs intended to deter the Soviet Union are those associated with attriting its military capabilities such that the Soviets cannot achieve their military objectives and cannot threaten US survival. Rather than using a punitive threat only to shape Soviet decisionmaking such that the Soviet leaders would not take highly provocative military actions, a balanced deterrent would promise to deny militarily the Soviet capability to achieve its politico-military objectives as part of the deterrent threat.

Assuming a Soviet defensive transition, the United States would no longer be capable of wielding, as an "ultimate deterrent," a threat to Soviet survival. Deterrent effect would flow from the high risk that would attend nuclear war, even in the context of thick strategic defenses, and the capability of the United States to survive attack and deny the Soviet Union its wartime objectives. The assumption supporting this type of deterrent is that the Soviet Union would be unwilling to accept the high risk of war if it could not anticipate achieving its so-called "theory of victory" requirements. As previously discussed, those requirements include "decisively" changing the "correlation of forces" during the initial phase of operations, and destroying the opponent's warmaking potential. A balanced deterrence should help ensure that the Soviet Union cannot anticipate meeting those requirements. Such a "denial of victory" deterrent should be particularly effective against the Soviet Union. John Erickson has noted:

> Soviet military opinion cautions against any 'adventurist strategy' which might prematurely initiate a total struggle, when the requisite 'correlation of forces' (sootnoshenie sil) cannot assure a favorable outcome (even 'victory') in such a struggle.[32]

The prospect of engaging in a long and risky unwinnable war, wherein the formidable US military and industrial potential would come into play, should be a deterring prospect indeed for Soviet leaders.

The primary difference between a denial and a punitive deterrent, beyond the differing concepts of what constitutes the deterrent threat, is the potential for damage-limitation. A punitive threat does not provide damage-limitation beyond the successful functioning of deterrence. A denial deterrent achieves damage-limitation through the successful functioning of deterrence and, if deterrence fails, by attriting Soviet offensive forces.

Some similarity in the denial and punitive approaches to deterrence exists. Most obviously, both assume an opponent will use a rational cost-benefit calculus in determining critical decisions. Consequently, to the extent that such an assumption is incorrect, or could be incorrect under certain conditions, both approaches to deterrence are weak. However, a denial deterrent makes provision for the failure of deterrence, a punitive deterrent does not.

An interesting question is why has the United States, in effect, eschewed the mission of defending the American homeland for many years? American strategic thought suggests that intercontinental means of delivery and nuclear weapons have rendered war "unthinkable" and separated the traditional linkage between policy and war, i.e., it is argued that war is no longer, as Clausewitz observed, an extension of politics by military means. Because nuclear weapons and their means of delivery have rendered nuclear war "unthinkable" according to the dominant American view, there is little need to think beyond the deterrence of war andplan to wage, survive, and recover from nuclear war should it occur. This dominant American view, that war had become unthinkable, is inconsistent with a serious commitment to strategic defense and damage-limitation.

The claim that nuclear war is unthinkable is only true if all nuclear-armed parties agree that it is true, and continue to agree that it is true—which, unfortunately, may well not be the case.

A certain change ushered in by nuclear weapons and long-range delivery systems is the immediate vulnerability of the US homeland to devastating attack. For Americans accustomed to the relatively cost free safety provided by the vast distances and oceans separating the United States from powerful potential foes it was, of course, easier to deny the existence of that new nuclear threat by asserting that war had become unthinkable than to make provisions for the survival of such a conflict. Obviously, the technical difficulties in achieving a damage limitation capability against strategic nuclear weapons discourage the undertaking.

CONCLUSIONS—NET ASSESSMENT

A net assessment concerning the approach to deterrence most appropriate for US needs must be   tentative. To conclude confidently that a particular approach to deterrence is superior to another presupposes that one can anticipate the line of reasoning and action that a relatively unknown Soviet leadership will pursue during some future (again unknown) crisis. Obviously, any such conclusion must be speculative. With that caveat firmly in mind it is possible to compare, logically, the three approaches to deterrence, and determine which is most appropriate for American needs.

US strategic forces are intended to deter massive and limited attacks on the United States, as well as extending deterrent coverage to allies and vital interests. All three of the deterrent approaches discussed above, MAD, "warfighting," and a balanced deterrent, should be effective in deterring massive attacks on the American homeland.

Regarding the punitive approaches to deterrence (i.e., MAD and "warfighting"), there should be little doubt about the credibility of the US deterrent threat to retaliate in the event the Soviet Union attacked massively. Assuming that the Soviet leadership behaves rationally, and believes that the cost of almost certain US nuclear retaliation outweighs any potential advantage of striking first, then either MAD or "warfighting" should provide an effective deterrence against a massive Soviet first strike. A balanced deterrent should also be effective against a massive Soviet first strike, again assuming Soviet rationality, and the Soviet perspective that no advantage is to be gained by initiating a very risky and unwinnable war.

However, the punitive approaches to deterrence should be less effective against more limited threats, such as a very limited counterforce attack against the United States or an attack against distant American allies and interests. As discussed above at some length, in the context of absolute US vulnerability, credibility should be low for a deterrent which threatens nuclear retaliation in response to Soviet behavior that does not entail a direct threat to US survival. Obviously, the closer Soviet behavior comes to threatening US survival (such as a limited nuclear attack on the United States as opposed to an attack on NATO Europe), the more credible should be the US strategic nuclear threat against the Soviet Union. However, for those potential Soviet actions that do not entail an obvious, immediate threat to US survival, punitive approaches to deterrence should provide a less credible deterrent than a balanced deterrent. A balanced deterrent would, in effect, reduce the apparent risk to the United States of extending its deterrence coverage to distant allies and interests—hence the credibility of that threat should be higher.

Nevertheless, it may well be that the deterrent benefits of a comprehensive defensive capability will not be available for several decades (or perhaps never if a defensive transition does not come to fruition for political or technical reasons). If so, the United States will have no choice other than to accept the risks associated with extending deterrence by punitive threat if it continues to accept extended deterrent responsibilities for its strategic nuclear forces. However, recognition that a balanced deterrent is more appropriate for meeting extended US deterrence responsibilities is a first and necessary step toward an adequate strategic nuclear policy.

Perhaps the most obvious advantage of a balanced deterrent is its potential for providing some security for the United States in the context of deterrence failure. There are, of course, no assurances that deterrence will function successfully during the forthcoming decades and punitive approaches to deterrence offer virtually no useful guidance or security in the event deterrence fails.

In summary, there can be no absolutely confident prediction that a particular type of strategic doctrine and strategy will provide a more effective deterrent against the Soviet Union. There is a lack of conclusive evidence upon which to base any such recommendation.

However, it is possible to conclude, logically, that a MAD-oriented punitive deterrent is inappropriate to support US deterrence responsibilities, and extremely dangerous in the event deterrence fails. A ''warfighting'' deterrence does not solve all of the inadequacies of MAD; it provides only a half-way solution if it does not incorporate a serious commitment to strategic defense.

Without defense, a "warfighting" position is as dangerous (if not more so) as MAD. It might, marginally, lower the threshold to nuclear use without providing for survival following that use.

A balanced deterrent, logically, should provide a more effective means for meeting the wide range of American strategic responsibilities. If a severe Soviet threat to America's European allies or other distant vital interest occurs during this decade, the American leadership is likely to discover just how unsuited the punitive approaches to deterrence are for extended deterrence. Perhaps more important, given the lack of surity that is possible concerning the functioning of deterrence, only a balanced denial approach can provide security for the American homeland in the event strategic deterrence fails. How likely is it that deterrence will fail or fail to apply to the next acute US-Soviet crisis (whatever and wherever it may be)? Again, it is virtually impossible to provide a confident assessment given the nature of the phenomena. That is perhaps the primary rationale for why the United States needs a nuclear doctrine and strategy consistent with a balanced approach to deterrence.

ENDNOTES

1. For a discussion of the "Schlesinger Doctrine," see Desmond Ball, "De ja Vu: The Return to Counterforce in the Nixon Administration" (Santa Monica, Calif.: California Seminar on Arms Control and Foreign Policy, December 1974); and Lynn Davis, *Limited Nuclear Options, Deterrence and the New American Doctrine,* Adelphi Papers, No. 121 (London: International Institute for Strategic Studies, 1976).

2. For a discussion of PD-59, see Harold Brown's presentation at the Naval War College, August 20, 1980, in Department of Defense, *News Release,* No. 344-80 (August 20,1980); US Congress, Senate, Committee on Foreign Relations, *Nuclear War Strategy,* Hearings, 96th Cong., 2d Sess., 1981; and Colin S. Gray, "Presidential Directive 59: Flawed but Useful," *Parameters,* vol. 11, no. 1 (March 1981), pp. 28-57.

3. For some public discussion of what is reported to be NSDD-13, see Robert Scheer, "Pentagon Aims at Victory in Nuclear War," *Los Angeles Times,* August 15, 1982, pp. 1, 2; and Michael Getler, "Administration's Nuclear War Policy Stance Still Murky," *Washington Post,* November 10, 1982, pp. 1, 2.

4. See, for example, Leon Sigal, "The Logic of Deterrence in Theory and Practice," *International Organization,* vol. 33, no. 4 (Autumn 1979), pp. 567-579.

5. For an excellent assessment of the Bishops' Letter, see Colin S. Gray, "Nuclear Deterrence and the Catholic Bishops," *Information Series,* No. 140 (Fairfax, Virg.: National Institute for Public Policy, April 1983); and Albert Wohlstetter, "Bishops, Statesmen, and Other Strategists on the Bombing of Innocents," *Commentary,* vol. 75, no. 6 (June 1983), pp. 15-35.

6. For a still useful examination of the entire spectrum of thought, see Robert Levine, *The Arms Debate* (Cambridge, Mass.: Harvard University Press, 1963).

7. For a detailed critique of MAD reasoning, see Fred Ikle, "Can Nuclear Deterrence Last Out the Century?" *Foreign Affairs,* vol. 51, no. 2 (January 1973), pp. 267-85; and Keith B. Payne, *Nuclear Deterrence in U.S.-Soviet Relations* (Boulder, Colo.: Westview Press, 1982), pp. 55-72.

8. As noted by Samuel P. Huntington, "The Renewal of Strategy," in *The Strategic Imperative: New Policies for American Strategy,* ed. Huntington (Cambridge, Mass.: Ballinger, 1982), p. 5.

9. *Ibid.*

10. Louis Morton, "Japan's Decision for War," in *Command Decisions,* ed. Kent Roberts Greenfield (Washington: US Government Printing Office, 1960), p. 124.

11. See, for example, Bernard Brodie, "The Development of Nuclear Strategy," *International Security,* vol. 2, no. 4 (Spring 1978). pp. 76-77.

12. As Thomas Schelling observed, "What local military forces can do, even against very superior forces, is to initiate this uncertain process of escalation. One does not have to be able to win a local military engagement to make the threat of it effective. Being able to lose a local war in a dangerous and provocative manner may make the risk—not the sure consequences, but the possibility of this act—outweigh the apparent gains to the other side." *Arms and Influence* (New Haven, Conn.: Yale University Press, 1966), p. 104.

13. For an early discussion of some "warfighting" notions, see *Limited Strategic War,* eds. Klaus Knorr and Thornton Read (Princeton, N.J.: Princeton University Press, 1962).

14. In US Congress, *Nuclear War Strategy,* p. 10.

15. See the discussion of LNOs by Dr. James Schlesinger in, US Congress, Senate, Committee on Foreign Relations, *U.S./U.S.S.R. Strategic Policies,* Hearings, 93rd Congress, 2d Sess., 1974, pp. 8, 12-13.

16. See Benjamin Lambeth, *Selective Nuclear Options in American and Soviet Strategic Policy,* R-2034-DDRE (Santa Monica, Calif.: RAND, December 1976).

17. See, for example, Wohlstetter, "Bishops, Statesmen, and Other Strategists On the Bombing of Innocents," pp. 5, 13.

18. See Keith B. Payne, "Deterrence, Arms Control, and U.S. Strategic Doctrine," *Orbis,* vol. 25, no. 3 (Fall 1981), p. 752.

19. Herman Kahn, *On Thermonuclear War* (Princeton, N.J.: Princeton University Press, 1961), pp. 132-133.

20. See Donald Brennan, "The Case for Population Defense," in *Why ABM? Policy Issues in the Missile Defense Controversy,* eds. Johan Holst and William Schneider (New York: Pergamon Press, 1969), p. 116.

21. Quoted in *Soviet Aerospace,* vol. 18, no. 16 (August 18, 1980), pp. 123-124.

22. For an elaboration of a balanced deterrent, see Colin S. Gray and Keith B. Payne, "Nuclear Strategy: Is There a Future?" *Washington Quarterly,* vol. 6, no. 3 (Summer 1983), pp. 55-66.

23. See N. Vasendin and N. Kuznetsov, "Modern Warfare and Surprise Attack," *Voyennaya mysl',* no. 6 (June 1968), translated FPD 0005/69, January 16, 1969, p. 46; S. Tyushkevich, "The Methodology of the Correlation of Forces," *Voyennaya mysl',* no. 6 (June 1969), translated FPD 0008/70, January 30, 1970, p. 26; I. Anureyev, "Determining the Correlation of Forces in Nuclear Weapons," *Voyennaya mysl',* no. 9 (June 1967), translated FPD 0112/68, July 11, 1968, pp. 35-45; and S. Ivanov, "Soviet Military Doctrine and Strategy," *Voyennaya mysl',* no. 5 (May 1969), translated FPD 0116/69, December 18, 1969, pp. 40-51.

24. See Helene Carsese D'Encausse, *Decline of an Empire: The Soviet Socialist Republics in Revolt,* (New York: Harper and Row, 1979).

25. See the text of President Reagan's March 23d speech in "President's Speech on Military Spending and a New Defense," *New York Times,* March 24, 1983, p. 20.

26. Barry Carter, "Let's Not Endanger Our Most Successful Arms Control Agreement," *Washington Post,* May 31, 1982, p. 17. The *Washington Post* has recently labeled the ABM Treaty "the jewel in the arms control crown." See "Mr. Reagan's New Defense Idea," *Washington Post,* March 25, 1983, p. 22.

27. See Dr. Patrick J. Friel's discussion of the technical difficulties of HEL BMD in "Space-Based Ballistic Missile Defense: An Overview of the Technical Issues," in *Laser Weapons in Space: Policy and Doctrine,* ed. Keith B. Payne (Boulder, Colo.: Westview Press, 1983), pp. 17-35; and Colin S. Gray, *American Military Space Policy* (Cambridge, Mass.: Abt Books, 1983), pp. 56-63.

28. For an examination of BMD for NCA and C^3I, see *A Strategic Transition: BMD for Critical NCA and C^3I Assets* (Fairfax, Virg.: National Institute for Public Policy, July 1983).

29. *Ibid.,* p. 62.

30. The cost of Soviet strategic defense activities increased from five times US outlays in 1970 to twenty-five times US outlays in 1979. See C.I.A., National Foreign Assessment Center, *Soviet and U.S. Defense Activities, 1970-1979: A Dollar Cost Comparison,* SR 80-10005 (January 1980), p. 9

31. See Keith B. Payne, "Transition to Strategic Defense: The Role of Ground Based BMD, and Arms Control," *Information Series,* no. 151 (Fairfax, Virg.: National Institute for Public Policy, July 1983).

32. John Erickson, "The Soviet View of Deterrence: A General Survey," *Survival,* vol. 24, no. 6 (November-December, 1982), p. 246.

CHAPTER 9

A NATO CONVENTIONAL RETALIATORY STRATEGY: STRATEGIC AND FORCE STRUCTURE IMPLICATIONS

by

Keith A. Dunn and William O. Staudenmaier

The United States is in the midst of a major strategic debate which will determine military strategy for the next quarter century. The US "Gang of Four" and others have called into question the entire basis of flexible response by arguing that the United States should adopt a policy of "no first use" of nuclear weapons, which of course challenges the US nuclear guarantee to NATO. Continental or coalition strategists like Ambassador Robert Komer are competing with maritime strategists like Secretary of Navy John Lehman and Francis J. "Bing" West, Jr. Big carrier navy advocates (Lehman, etc.) are opposed by naval strategists who favor smaller carriers and cruise missile technology (Admirals

Stansfield Turner, Elmo Zumwalt, Worth Bagley, etc.). Military Reform Caucus supporters believe that the United States emphasizes attrition warfare too much and does not pay enough attention to maneuver warfare. Moreover, they contend that the United States is too infatuated with technology and needs to buy smaller less expensive equipment for all the Services to support maneuver warfare concepts. A disparate group of people are divided over the issue of whether the United States should adopt an offensive or defensive military strategy. Today, it appears that the offensive is in the ascendancy.

Samuel P. Huntington's belief that NATO should adopt a conventional retaliatory strategy is one of the recent cases made for an offensive operational strategy which has attracted much attention within the US and European defense community. His arguments are outlined in Chapter 2. Using Huntington's concept as their starting point, each of the other contributors to this volume have addressed several basic issues associated with defense and deterrence in the 1980s. Each author attempts to answer a basic question: how much emphasis should the United States place upon defense versus deterrence in the future? The preceding chapters speak for themselves. Therefore, this chapter will focus on the policy, strategy, and force structure implications associated with the Huntington proposals.

THE STRATEGIC ENVIRONMENT

For the past thirty-five years, US military strategy has been based upon a particular set of strategic concepts. These concepts include: nuclear deterrence (based upon mutual vulnerability and assured destruction), escalation control, superpower conflict avoidance, collective defense, conflict control (i.e., a desire to limit the geographic scope, intensity, and duration of any conflict), and crisis management. They were developed when the United States was the world's preeminent nuclear power and had a clear technical, qualitative advantage in theater and battlefield nuclear weapons. In addition, these concepts and the resulting military strategies were developed when US alliances were relatively cohesive and the United States was believed to be not only the military, but also the political and economic leader within the Western world.

Events in the 1970s and 1980s clearly demonstrate that this strategic environment has changed. First, the United States no longer has strategic nuclear superiority. Since at least the mid-1970s, a rough equivalence in strategic nuclear weapons has existed between the two superpowers, although significant asymmetries in particular delivery systems exist. Second, Soviet conventional military capabilities have improved. The USSR is militarily involved in areas of the world where it never has been previously. The Soviet Union still is inferior to the United States in its conventional global power projection capabilities; however, its exploitation of events in Angola, Ethiopia, and Afghanistan in the mid to late 1970s contrasts sharply with its ineffective attempts to use its military forces for political purposes in the Congo in the 1960s and Peru in 1970. For at least the remainder of the decade and possibly for most of the 1990s, Soviet power projection capabilities will continue to be constrained because of three traditional limiting factors: (1) an imbalance between naval logistical support ships and combatants that limit how long Soviet combatants can be deployed at sea; (2) a mechanized and tank heavy Army which gives the USSR certain advantages in the European theater, but is very difficult to deploy, sustain, and support in theaters noncontiguous to the USSR; and (3) a lack of reliable friends and allies who want to see Soviet military presence consolidated within their countries. Nevertheless, the simple fact that Soviet ships, planes, or advisers may be in a particular country or deployed in a theater on a continuous basis significantly changes the level of risk that today's political decisionmaker must be willing to accept before committing the US military to action. As a result, even though it is possible to demonstrate that Moscow lacks the capability to support and sustain conventional military operations in many distant theaters like Latin America, Africa, Southeast Asia, or the Pacific, except against very unsophisticated adversaries, Soviet global conventional deployments have altered the way that many American military strategists and policymakers view the strategic environment. This perception alone is significant because it is possible some of the more optimistic assessments of Soviet power projection capabilities could result in deterring the United States in the future when actual Soviet capabilities may not warrant such US restraint.

Third, US alliances are possibly in more disarray than at any time in the past. In part, this is because the United States is no longer the sole political and economic leader of the Western world. The growing economic strength of Western Europe, Japan, and the oil-rich nations of the Persian Gulf/Middle East have caused those nations to pursue not only more assertive independent economic policies, but also to challenge the United States in the international political arena.

The current weaknesses in the US alliance system also result from different perceptions among sovereign nations over what are the best ways to obtain competing national interests. Simply put, our allies do not always perceive the threat in the same way that the United States does, because their interests and objectives are different from ours. This does not mean, as one author has stated, "that American and allied [NATO] perceptions of the Soviet Union are so different that one must ask whether Washington and the allies continue to agree that there is in fact a Soviet threat to the West."[1] NATO allies—and other allies as well—recognize a Soviet threat. In recent years, however, they have been less inclined to accept unquestioningly that the American response is the right or only way to react. Monolithic communism, which was the basis of the US post-World War II military alliance system (NATO, CENTO, SEATO, and sundry bilateral agreements), no longer has the credibility that it did in the 1950s or even the 1960s. As a result, alliance consensus is difficult to forge and fragile to maintain.

Fourth, Soviet-American relations have severely regressed. The early 1970s were a period of European confidence in US-USSR relations and were the high point of detente, arms control, superpower crisis management, bilateral trade, and diplomatic, cultural, and educational exchanges. Steadily, beginning with Angola, Ethiopia, and then Afghanistan, Soviet-American relations have soured. As the two superpowers entered the 1980s, their bilateral relations were, according to the International Institute for Strategic Studies, "more strained than at any time since the death of Stalin in 1953."[2] Desires for mutual cooperation have given way to suspicions, irritations, and recriminations strongly resembling the worst of the Cold War period. For the moment, at least, ideology has replaced pragmatism between the superpowers.

Finally, the disintegration of the post-World War II bipolar environment has caused not only a more diversified and fragmented world, but also a more interdependent one. A primary result of the general diffusion of economic, political, and military power that has occurred over the last thirty-five years is the inability of the superpowers to control and shape events to the degree that they would like. This phenomenon has been amply exhibited in the volatile and important region of Southwest Asia in recent years. Despite its efforts, the United States was unable to curb the indigenous uprising that led to the fall of the Shah of Iran. Likewise, the Soviet Union lacked the ability in Egypt or Somalia to keep those strategically important countries from expelling Soviet forces. While neither superpower approves of nuclear proliferation, it continues to occur in Southwest Asia even though the participants—Pakistan and Iraq—are ostensibly "friends and allies" respectively of the United States and the Soviet Union. Washington's and Moscow's inability to influence the protracted war of attrition between Iran and Iraq, although neither superpower's interests are advanced by the conflict, is just another example of how difficult it is for the superpowers to control events, even when they are occurring in a strategically important region.

Moscow and Washington are not impotent. However, in a strategic environment where the primary causes of recent regional conflicts are more often than not indigenous with historical roots that have little relationship to East-West competition, where both superpowers are dependent upon raw materials and products from the politically unstable Third World, and where alternative arms suppliers and a growing indigenous Third World arms production capability exist, traditional superpower crisis management tools (e.g., show of force, increasing or cutting off arms, etc.) are becoming less effective. Not only do nations have places other than the United States or the USSR to obtain military and economic assistance, but also the web of global interdependence makes strange bedfellows as witnessed by the fact that adversaries like Israel and Syria have provided arms and other assistance to Iran in the Iran-Iraq War.

Reaction to Change. Most analysts can agree that the strategic environment has changed. Many can even agree on how it has changed and what are the salient characteristics of the new environment. However, it is at this point where the consensus begins to unravel. There is virtually no agreement for how the United States should react or what solutions it should attempt to

191

cope with the new strategic environment. For example, Huntington has argued that the traditional US strategic concepts "are of dubious relevance to the conditions of the 1980s."[3] Other critics of current US military strategy argue that the global strategic center of gravity has shifted from Europe and Northeast Asia to the Persian Gulf and the other resource-rich Third World nations requiring a refocus of US military strategy and forces toward those regions.[4] There is yet another group of strategists who recognize the importance of the Third World, but believe as does Ambassador Komer that an overemphasis upon preparing and sizing US forces for lesser contingencies in the Third World makes the United States a prisoner of the "fallacy of the most likely argument." In other words, the United States may acquire the capability to deal with Grenada "liberations," Iran hostage raids, or countercoup attempts in a myriad of countries but, as a result, fail adequately to prepare for the defense of what is most vital to the United States—the European political, economic, and military heartland. This latter group believes that US military strategy and its underlying concepts are basically sound. Therefore, they support making US strategy more effective through better alliance cooperation rather than changing it.[5]

Has the strategic environment changed so much that the United States should revamp its military strategy? Obviously, there is no easy answer to this fundamental question. Otherwise, the heated strategic debate would have cooled long ago. Clearly, some things have changed (e.g., strategic nuclear equality, improved Soviet power projection capabilities, interdependence, power diffusion, etc.). However, many of the so-called dramatic changes were inevitable, are irreversible and, in fact, were fostered by the US policy. For example, the United States sponsored, supported, and encouraged West European and Japanese economic revitalization after World War II in the belief that economically prosperous nations were more politically stable and that this best served US interests and objectives. In the short term, the relative decline in US economic and political status is disconcerting. But it would be difficult to support an argument that the United States would be better off today if Europe and Japan were as economically weak and vulnerable as they were in the 1950s. The only real opportunities that the USSR has been able to exploit in the post-World War II period have been in countries which were weak, isolated, and vulnerable. Europe and Japan may not follow US leadership to the extent some would prefer, but they clearly do not want Soviet political power consolidated either in their region or anywhere else. In that regard, not only has the strategic environment not changed, but also two of the world's economic power centers are better able to resist the USSR which helps the United States obtain an important objective—containing the spread of Soviet influence and power.

Conflict Avoidance. Another area where no major change to the strategic environment has occurred is in the area of superpower conflict avoidance. Since the end of World War II—even before Moscow obtained nuclear weapons—a tacit understanding between the two superpowers existed: namely, direct Soviet-US conflict had the possibility of escalating out of control—a situation which neither Washington nor Moscow preferred. In recent years, an argument has been advanced that obtaining strategic nuclear parity and improving its conventional power has made Moscow more brazen and willing to risk confrontation with the United States. The empirical evidence to prove such assertions, however, is clearly lacking.

The Brookings Institution has produced the most comprehensive study to date on the use of Soviet military power for political purposes. This study concludes that Moscow has employed its military forces "neither recklessly nor clumsily, but with prudence and sensitivity."[6] If the United States has been unwilling to aid an ally or friendly nation that is being pressured or coerced by the USSR or one of its allies, Moscow has not always showed similar restraint. But rather than becoming more brazen and aggressive in its employment of forces, Moscow has employed its forces pragmatically seeming to be careful to insure that its actions would result in minimum risk to the Soviet state and its interests. Moscow has been sensitive to situations where its forces might come into direct military conflict with the United States, particularly where the vital interests of one or both superpowers were involved.

While the Brookings study was completed before the Afghanistan invasion, there have been no significant changes that would cause one to believe that Moscow's behavior has altered significantly from its traditional path.[7] Given America's historic lack of interest not only in Afghanistan, but also in South Asia generally, Soviet planners, from a military perspective at least, correctly concluded that there was very little chance that an invasion of Afghanistan would lead to direct superpower conflict. American and Soviet interests in Afghanistan were too asymmetrical for the invasion to lead to military conflict.

The principle of superpower conflict avoidance still seems to affect Soviet policy. Of course, an increase in Soviet conventional capabilities complicates the situation because the United States can no longer assume that it can deploy forces to crisis areas with impunity or not expect to encounter Soviet conventional forces. The possibility of accidental conflict in Third World areas where the nature and extent of either superpower's interests are ill-defined exists when both nations have the capability to send forces to signal resolve or interest. Nevertheless, Soviet actions and words indicate that superpower conflict avoidance—because of the possibility of nuclear war—continues to be a major aspect of Soviet policy. According to Dimitri Simes:

> Traces of the old nuclear war-winning school of thought may be found in some Soviet military writings. Neverthless, an assessment of Soviet literature strongly suggests that during the last 15 to 20 years, the Soviet leadership has gradually but steadily, albeit with zigzags, moved toward acceptance of mutual deterrence and a rejection of prior claims that the Soviet Union possesses superior forces capable of assuring victory in a nuclear war.[8]

Brezhnev regularly warned that because nuclear weapons existed, the USSR would have to prepare for the possibility of nuclear war. However, less than six months before his death, he reasserted his belief that no one would be a winner of a nuclear war:

> We are convinced that no contradictions between states or groups of states, no difference in social systems, ways of life or ideologies and no transient interests can eclipse the fundamental need for all the peoples—the need to safeguard peace and avert a nuclear war . . . because should a nuclear war start, it could mean the destruction of human civilization and perhaps the end of life itself on earth.[9]

Since Brezhnev's death, there have been no major changes in the Soviet position. As Lawrence Caldwell and Robert Legvold have recently argued, there has been increased talk within Soviet elite circles over the last year about the dangers of war.[10] This in its own right should be a troubling phenomenon for the West to observe because it reflects a less than optimistic view about superpower trends and relations. There may be a greater tendency now for the post-Brezhnev leadership to lecture the United States almost bitterly about American irresponsibility. But, the current Soviet leadership—like Brezhnev—seems to recognize that a strategic nuclear war would serve no useful purpose. These attitudes are summed up in a March 1983 interview that Yuri Andropov had with *Pravda:*

> The US Administration continues to tread an extremely perilous path. The issues of war and peace must not be treated so flippantly. All attempts at achieving military superiority over the USSR are futile. The Soviet Union will never allow them to succeed. It will never be caught defenceless by any threat. Let there be no mistake about this in Washington. It is time they stopped devising one option after another in search of the best ways of unleashing nuclear war in the hope of winning it. Engaging in this is not just irresponsible, it is insane. . . .
>
> Today all efforts must be directed towards one goal, that of averting nuclear catastrophe.[11]

DETERRENCE IN AN ERA
OF STRATEGIC NUCLEAR PARITY

If there is one issue on which most recent critics of US military strategy seem to agree, it is the belief that the US capability to support extended deterrence—i.e., the protection of allies and friends that cannot on their own deter the USSR—has been seriously undermined because the United States no longer has strategic nuclear superiority. Particularly, critics argue that the loss of strategic nuclear superiority makes NATO's strategy of flexible response a questionable military strategy since the United States no longer has escalation dominance.[12]

The belief that the US nuclear deterrent threat has been largely stalemated has led a new group of revisionist strategists to suggest that the United States should adopt new methods for reinforcing deterrence. Horizontal escalation and a revived interest in offensive land strategies for Europe are two of the latest methods which have received a great amount of interest in the national security community in recent years. These two methods should be examined in terms of two questions. First, how well do they achieve US interests and objectives? Second, what impact do they have on escalation? In the era of strategic nuclear parity, when both superpowers have the capability to destroy each other, control is the essence of strategy.

Horizontal Escalation. Simply stated, horizontal escalation is based on the belief that if the United States is incapable of defeating Soviet aggression at the primary point of attack, it should attempt to open a second front or attack some valued Soviet possession. This, it is argued, will do two things. First, expanding the conflict will exacerbate Soviet traditional fears of a two-front war. Prior to conflict, such a strategy is expected to deter the USSR from aggression. If deterrence fails, an operational strategy based upon horizontal escalation, it is argued, will drain Soviet forces and not allow the Kremlin to concentrate or mass its forces for victory. Second, horizontal escalation, advocates claim, will provide an offensive character to US strategy and increase US freedom of action.

Horizontal escalation, particularly with a maritime focus, is closely identified with the Reagan Administration. However, different versions of this concept have weaved their way through other administrations. According to Henry Kissinger, the primary reason that the Nixon Administration opened its arms to the People's Republic of China (PRC) was to create concerns in the Kremlin regarding two-front war possibilities.[13] The Carter Administration turned toward China and hinted that it would be interested in selling weapons to the PRC for the same reasons. Following the fall of the Shah of Iran and the Soviet invasion of Afghanistan, the Carter Administration considerd horizontal escalation alternatives because it lacked the capabilities needed to implement the Carter Doctrine in Southwest Asia.[14] Nevertheless, the Reagan Administration continues to be most identified with this concept for good reasons. Some individuals who joined the administration in influential policy positions were already considered advocates of horizontal escalation.[15] Also, during 1981 and 1982, official administration announcements, including the Secretary of Defense's annual posture statement, indorsed the concept as official policy.[16]

Since 1983, Reagan Administration representatives in the Department of Defense (DOD) have made fewer direct references to horizontal escalation and, at least, one advocate of the concept—Francis J. West, Jr.—has left the administration. These events have caused some observers to argue that horizontal escalation has been rejected. The evidence, however, does not support such a conclusion. To properly evaluate the role of horizontal escalation in the administration's strategy requires an understanding of two important national security concepts. The differences between declaratory policy and acquisition policy must be considered.

Declaratory strategy or policy is the publicly proclaimed rationale for a particular government action, often associated with military programs. Acquisition policy, on the other hand, is the procurement action that a nation takes in the development of its military force structure. It is, of course, easier to change declaratory policy than it is to change the direction of acquisition policy because of the extensive lead times associated with modern military procurement.

For example, the declaratory strategy relative to horizontal escalation has changed over the last two DOD annual reports. The Reagan Administration has sought to distance itself from a firm statement in support of horizontal escalation, much like the Ford Administration when it expunged detente from its lexicon in 1974 when the term became a political liability. Phrases invoking images of horizontal escalation such as "we must recognize that, in a conventional war, in a region like Southwest Asia, the geographic limits of combat cannot be taken for granted" and "the requirements for maritime access to that region may well require us to respond to naval attacks not necessarily limited to the geographic boundaries of that region" are found in the FY 84 DOD Annual Report to Congress and in other public statements by Administration representatives.[17] They are essentially statements of horizontal escalation, if phrased in more palatable terms.

Phrases that would even suggest that horizontal escalation was still a viable concept, however, do not appear in the FY 85 DOD Annual Report.[18] Therefore, horizontal escalation as a declaratory strategy is dead. Paradoxically, when one considers acquisition policy, it is clear the concept is alive and well—at least in terms of the military capabilities needed to carry it out. The maritime naval capabilities—600 ships and fifteen aircraft carriers—needed to perform simultaneous operations in multitheaters (the indispensable requirement of horizontal escalation) remain a major element of the Reagan defense program. While horizontal

197

escalation may no longer be the declaratory strategy of the United States, it does remain as US acquisition policy.

A further reason why some analysts still believe that horizontal escalation is no longer valid exists because of a basic misunderstanding of two types of planning that occur in the military community: force planning and operational planning. Force planning develops military forces and capabilities for the midterm (three to ten years in the future). The force planning system is primarily the responsibility of the Department of Defense and the military Services—Army, Navy, and Air Force. The output of this system is the allocation of resources to build future US military forces. Operational planning, on the other hand, focuses on current or near-term (one to two years) forces and threats. It deals with planning designed to employ existing military forces to achieve US interests and objectives. Primarily, operational planning falls in the purview of the Joint Chiefs of Staff and the Unified and Specified Commanders.

It is important not to confuse the two systems when evaluting the effectiveness of military strategies. The operational plans that the Joint Chiefs of Staff and the Unified Commanders develop to support contingencies, for example, in Southwest Asia are based on existing capabilities. It is the way we would fight a war in that region should it occur tomorrow. War plans developed in the operational planning system encompass plausible contingencies ranging from a coup attempt against a friendly Gulf State to a Soviet invasion of Iran. These plans represent the way we must fight because of operational shortcomings, not necessarily the way we would prefer to fight.

The way we would hope to fight in the future is within the purview of the force planning system. It develops plans to build the military force structure for the way that we would like to fight the war in Southwest Asia, for instance. Since it takes time to build this force structure, the force planning system deals with the threat, strategies, and force structure of the future. The problem comes— and it is a common problem—when defense analysts confuse the two. The initial US debate that occurred from 1979-80 as the Rapid Deployment Joint Task Force (RDJTF) was created is a case in point. Many defense analysts criticized the military for developing plans for a nonexistent RDJTF. The critics were applying operational planning criteria to a force planning concept.

Even the most ardent supporters of horizontal escalation realize that the United States lacks the current capabilities to execute that strategy now. Those who support horizontal escalation are really thinking about force planning, whether they realize it or not. In other words, they advocate that the United States—over the next three to ten years—build the capabilities to be able to perform horizontal escalation. Since the Secretary of Defense's annual report to Congress has more to do with the force planning cycle than the operational planning cycle (although it has applicability to both), one can argue that horizontal escalation—even though the language has been purged from the administration lexicon—is still alive and well.

Until recently, conventional horizontal escalation options have had a maritime character. However, the offensive military strategies for Europe that Huntington and others have offered in recent months are essentially horizontal escalation concepts using land, rather than naval forces. Their purpose is to outmaneuver the USSR on the European battlefield, rather than by the seas, through a form of strategic jujitsu. According to these new concepts, NATO will initiate offensive actions into Eastern Europe. There has always been an offensive element to NATO's military strategy (e.g., air interdiction) and the current Supreme Allied Commander in Europe, General Bernard W. Rogers, has suggested improving the capability with some of the new conventional weapons that should be available to the alliance before the end of the decade.[19] The difference between NATO's military plans and the new offensive strategies is, however, more than style. The primary purpose of fighting the "deep battle," as it is referred to now is to stabilize the battlefield and to keep Soviet strategic echelons from reinforcing Warsaw Pact forces that are in contact. The primary objective then is military. Huntington's proposed new offensive land strategy's primary purpose, however, is political. Threatening to attack into Moscow's *cordon sanitare* if warfare should develop in Europe, it is argued, will enhance deterrence by placing at risk something that is politically valuable to the USSR.

Problems with Horizontal Escalation. Three generic problems are associated with horizontal escalation whether it is executed with sea or land forces. A more specific problem is also associated with European land versions of horizontal escalation which should be discussed—if the United States could acquire the military forces to

199

execute a NATO offensive military strategy, would it even be necessary to change NATO's military strategy? The generic problems will be discussed first.

The first problem is that both land and maritime variants of horizontal escalation imply an escalation of objectives from the limited goal of countering regional Soviet aggression to effecting the ultimate military defeat of the USSR. Maybe the United States should adopt such a policy. However, that is a political and not a military question. The military strategist's job is to construct military strategies which attempt to achieve US objectives within the parameters of policy guidance. Superpower conflict avoidance, because of the fear of uncontrollable escalation, continues to be a policy constraint within the United States (and the Soviet Union for that matter). As a result, alternatives like horizontal escalation which suggest that the United States should expand a conflict because of an inability to meet aggression at some particular point not only increase the risk of nuclear escalation by expanding the points of friction between the superpowers, but also run counter to over thirty-five years of successful policy guidance: no one is the winner if the superpowers directly confront each other with military forces because of the risk of nuclear escalation.

Second, in the United States, military strategy must achieve some political objective. Two important US political objectives are defense of the territorial integrity of NATO nations and insuring the flow of Persian Gulf oil to the United States and its industrial allies. Supporters of both maritime and land oriented horizontal escalation strategies generally support these two objectives. However, the argument that, if the Soviet Union should move militarily to cut off Persian Gulf oil and the United States cannot prevent it, the US forces should seize Angola, Cuba, or some other Soviet strategic outpost does not fulfill the primary objective. Seizing Soviet overseas possessions may fulfill needs to punish the USSR. However, strategy must accomplish more than retribution. Seizing Soviet outposts might deter the Soviet Union from committing aggression in the future, but it would not contribute to removing Soviet forces from the Persian Gulf oil fields. Therefore, as long as American objectives in the Persian Gulf are defined as keeping the oil flowing over the long term and insuring the region does not fall under the domination of hostile outside powers, there can be no balance of a strategic equation that justified the loss of

Persian Gulf oil because America has seized a Soviet outpost, no matter how important it is to the USSR.

The same principle applies to retaliatory offensive strategies in Europe. Seizing Jena and Leipzig or even Warsaw or Prague cannot justify the loss of significant European territory unless NATO is willing to change the political underpinnings of the alliance. What NATO would get in return would not be offset by the important economic, political, social, and cultural loss the Western World would suffer.

A third generic problem relates to conflict termination. For the sake of argument, assume that the Soviet Union attacked Iran and a US counterattack at Soviet naval bases on the Kola Peninsula or the Far East maritime provinces engaged enough Soviet forces to stabilize the Iranian front before the USSR occupied the oil fields near the Persian Gulf. What happens then? Could the United States or the Soviet Union negotiate a settlement while American and Soviet military forces are engaged in an area of vital Soviet interests and Soviet territorial integrity is under attack? Would the USSR now have to be defeated on the original secondary front? Or assume that a successful strategic maneuver into Eastern Europe forced Moscow to stop its attack on NATO before all of West Germany was occupied. How do the superpowers politically negotiate their way to the *status quo ante* when in practical terms what Moscow holds is more important to NATO than what NATO forces have occupied in Eastern Europe? NATO would not be negotiating from a position of strength in this latter situation. These are the sorts of strategic questions that must be asked of those who favor maritime or land forms of horizontal escalation before the United States and NATO reject traditional strategic concepts.

The more specific problem associated with land versions of horizontal escalation is that advocates have begged the important question: how many additional forces would be needed to support their proposals? The maritime force structure requirements for horizontal escalation are clear. The official maritime position is that a total of fifteen operational US carrier battle groups, if supported by allied navies, would be sufficient. On the other hand, those who support army or land versions have not presented their numbers in such a clear fashion. A safe assumption, however, is that the United States would require more than five divisional equivalents, two armored cavalry regiments, and one infantry brigade in Berlin that are now stationed in West Germany. Determining how many additional land forces would be required to launch an offensive into Eastern Europe is difficult, however, and possibly that is why advocates have avoided specifying an exact number. Nevertheless, it is possible to suggest that four major preconditions would have to be met before a European land version of horizontal escalation could be successful. First, as Huntington has argued, any extra divisions would have to be on the ground in Europe ready to fight before D-Day. In the case of American divisions, it would be preferable if they were actually stationed in Europe. Second, if reinforcements are believed to be necessary now, they would be just as important when NATO was trying to defend one sector while other forces attacked into Eastern Europe. Third, US forces earmarked for the conventional retaliatory action could not be siphoned off to non-European theaters if the United States was faced with simultaneous conflict. Fourth, and most important, the force that maneuvers into Eastern Europe not only must be capable of defeating the Soviet, Polish, Czech, or East German forces that it faces immediately to its front, but also it must be able to survive an attack by the Soviet strategic second echelon deploying from the Western Military Districts. If the maneuver force cannot survive the second strategic echelon, it will be an irritant, but not a decisive use of military power.

The USSR has at least sixty-five divisions (twenty-three tank, thirty-seven motorized rifle, and five airborne divisions) in its European Military Districts which could be dedicated to its strategic second echelon. Furthermore, there are an additional sixteen divisions from the Moscow, Ural, and Volga Military Districts which could form a strategic reserve.[20] Assuming a best case for the NATO force, let us estimate that Moscow uses only its twenty-four divisions from the Baltic and Byelorussian military districts as its strategic second echelon. (If the tank and motorized rifle divisions from the Carpathian Military District were included, this echelon would be thirty-six divisions). Given the Warsaw Pact forces already in Eastern Europe, it would require at least ten to fifteen additional US divisions to carry out this operational plan.[21] Not only would the force have to maneuver and fight its way into Eastern Europe, but also it would have to defend itself from the inevitable counterattack of the Soviet second strategic echelon. In addition, it would have to defend a long and vulnerable supply line through hostile territory as well as a 360 degree defense perimeter. Each of these tasks is a difficult undertaking, particularly when the risk of nuclear war would be high. Trying to do all of them simultaneously is virtually impossible under current circumstances.

Again, however, for the sake of argument let us assume that NATO could satisfy all of the preconditions and obtain a maneuver force with the necessary associated tactical air support and strategic lift. If this were possible, why would NATO want to change its strategy? With an in-being force this large, European defense and deterrence requirements would be guaranteed.

Deterrence is based on the ability of one nation to convince another nation that the risk associated with taking a particular action is not worth the possible benefit or punishment. Deterrence is enhanced also if a potential aggressor cannot calculate with reasonable assurance what costs it will have to incur for taking some action. Of all military theaters, Europe fulfills the most important ingredients for deterrence. The potential of uncontrollable escalation is great and both sides seem to recognize this fact. Each superpower has vital interests in the region. Each has substantial conventional and nuclear forces within the theater. The risk of either superpower attempting to use military force to seize the other's vital interests is nuclear war. While the United States may prefer that more cohesion exist within NATO and that there were more conventional forces within the region, Moscow traditionally has perceived a more unified NATO than have Western observers. Like Western military planners, the USSR is very adept at worst-case planning. For all of these reasons, it would be strategically unwise to jeopardize the tenuous consensus that ties sixteen sovereign nations together by adopting new strategies that are not supported by our European allies. Besides, if it were possible to muster the political will within NATO to finance the type of force structure increases required by some of the new alternatives, there would be no need to apply them in any case.

MILITARY CONSEQUENCES

To this point this analysis primarily has focused on national strategy and policy implications of changing NATO's military strategy for good reason. Most of the current critics, particularly Huntington with his conventional retaliatory strategy, are fundamentally arguing that significant changes in the strategic environment require changes in US policies. They are not particularly concerned with the military implications of those policy changes. Huntington, at least by implication, leaves the specific military force structure changes needed to execute his policy changes rightfully to military professionals.

This approach is clearly consistent with the US strategic process where the civilian leadership is dominant. US strategy, including setting interests and objectives, is established by the elected political leadership and their designated civilian representatives. Policy constraints—the risks that the United States is willing to accept to achieve its interests and objectives—in theory, at least, if not in practice, are developed also by the political leadership. Military advisers contribute by providing their best military judgment concerning the military suitability (i.e., the desired military strategy will achieve some political purpose), acceptability (the military objective can be obtained within reasonable financial costs), and feasibility (the military strategy has a reasonable chance of success) of the declared political policy and military strategy. However, to complete the strategic process, we must examine the military consequences associated with the Huntington proposal for two reasons. First, the Huntington idea was used as the organizing theme for this book and, second, the problems associated with the Huntington proposal equally apply to most other land versions of horizontal escalation.

Presuming that additional forces could be acquired, the most immediate issue that a military strategist would have to confront before advising his political superiors about the military suitability, acceptability, or feasibility of a retaliatory strategy is the stationing of the additional forces. The military adviser would need a political decision on whether the American forces would be stationed in Europe or remain in the United States to be deployed prior to D-Day. Political decisionmakers would have to consider different military factors with each stationing option. If the US element were to be permanently stationed in Europe, where would it be located? Presumably it would have to be on or near the Inter-German Border to be able to initiate a conventional retaliatory strike, as Huntington has defined the operation. However, practically speaking, there is little spare space in West Germany for additional peacetime deployments. Currently the kasernes are not adequate to meet US needs. Moreover, US forces already lack maneuver room and training areas for existing forces. Spreading US divisions across other Corps areas would not be a suitable solution. Other allied Corps face the same type of problem as do the US V and VII Corps. More importantly, linearly deploying the additional forces would defeat the primary objective of increasing the force structure in Europe, which is to be quickly massed for a retaliatory strike in Eastern Europe. There would also be a financial, balance of payment, problem associated with stationing significantly more US divisions in Europe.

If the decision were to station the US element of NATO's retaliatory force in the United States, the miltiary advisor would have to inform his political superiors that he would need at least three political commitments before he could say with any degree of assurance that the strategy and policy possibly could succeed. First, sufficient sea and airlift capabilities would have to be purchased simultaneously as the maneuver forces were procured.[22] No useful purpose is served in buying a conventional retaliatory force but not giving it the strategic lift required to carry out its mission. In fact, without the appropriate strategic mobility assets, money could be more wisely spent in other ways. Second, policymakers would have to avoid any temptation to spread the CONUS (continental United States) stationed force throughout the United States among different congressional districts. To reduce staging and deployment times, the divisions would have to be stationed in the eastern part of the United States, either collocated with a major Air Force base or very near a major port with equipment possibly up-loaded. Third, political decisionmakers would have to agree to deploy the force even if strategic warning were ambiguous. If the United States waited for unambiguous strategic warning—something NATO would probably never get in any event—the retaliatory force may never get to Europe in time, a problem that US forces already face. However, if the strategy and policy are to attack quickly into Eastern Europe rather than counterattack to drive Soviet forces from any NATO territory that is occupied by the Warsaw Pact, getting to Europe before hostilities actually begin becomes a critical requirement that military planners would have to advise political leaders about. The risk of acting on ambiguous warning is that the United States could actually escalate a crisis into a conflict—the World War I syndrome. Nevertheless, the policymakers would have to accept this risk if a conventional retaliatory strategy with a CONUS stationing was chosen. Prepositioning division sets of equipment (POMCUS) could reduce some of these problems. However, American forces still would have to get to Europe and draw their equipment before the conflict began. Some analysts question if it would be possible to draw equipment from current POMCUS sites during a conflict given the fog of war and problems of battlefield congestion not to mention possible Soviet air, missile or chemical attack.[23]

The costs associated with new divisions also would be significant. According to the Congressional Budget Office, the five-year cost for acquiring just four divisions stationed in the United States would be $37.8 billion.[24] It would cost approximately $7.6 billion to buy equipment, $9.1 billion to open and operate four new bases, $9.5 billion to operate the divisions, and at least $11.6 billion for manpower.

Military strategists would also have to advise their political superiors that the possibility of reinstituting the draft would have to be given serious consideration to make the new strategy work. Without a draft, the strategy may not be militarily feasible. Each new division would require approximately 18,000 personnel. This accounts only for the basic division and does not consider the larger division slice which in terms of procuring personnel could push the total numbers of new US recruits needed as high as 480,000 if the US contribution to NATO's retaliatory offensive were ten divisions. With the decreasing US manpower pool projected through the turn of the century, a draft or some sort of universal military service may be the only possible way for the United States to acquire the necessary personnel. Moreover, the pay scales associated with the current army would probably force the military to recommend a return to the draft in order to reduce financial costs.

Finally, executing an offensive operational strategy into Eastern Europe would require heavy armored and mechanized US forces, as Huntington points out. The US Army's focus in the last four to five years, however, has been to "lighten" its forces to make it better able to respond to the variety of lesser contingency threats that policymakers believe the United States will face in the future.

Obviously, the US Army can reverse its direction. This has happened in the past. But, the real issue is should it change its direction by creating forces that are focused primarily on a single European scenario or should it continue to try and hedge by preparing for other contingencies where heavy forces are not as useful? If the political decisionmakers choose the former, the military strategist must forewarn them that this decision may limit the Army's ability to respond to numerous Third World contingencies in Latin America, Africa, Southwest Asia, and the Pacific.

CONCLUSION

In recent years, the emerging reality of superpower strategic nuclear parity has strengthened the belief that only conventional military forces have political utility. This belief has led strategists to seek ways to break the nuclear stalemate by ending their long-term neglect of military operational issues. Professor Huntington's proposal for a conventional retaliatory offensive into Eastern Europe as a means to enhance deterrence is squarely in the mainstream of this effort.

This proposal, bold in its conception and appealing in its emphasis on offensive action, however, is flawed in its political and operational dimensions. Politically, it appears to have little chance for adoption because it runs counter to the political and psychological need of the Europeans to believe in the American nuclear guarantee. To assert that extended deterrence is no longer functionally important runs counter to events unfolding in Western Europe. The implementation of an offensive retaliatory strategy is unacceptable to most West European governments. Even the Soviet perception of the credibility of such a strategy is low because Soviet planners believe they would be able to counter such an allied incursion into Eastern Europe with relative ease.

Operationally, the strategy would be difficult for NATO to implement. First, raising the required additional divisions could cause the United States to reinstitute the draft and the forward stationing of these new forces would strain American defense budgets and European political relations. If the new divisions were formed from European manpower assets, the political and financial costs would probably not be acceptable to continental governments. Second, many Atlantic strategists have pointed out the enormous difficulties facing anyone who would change alliance strategy. The agreements are the result of compromises so fragile that even the suggestion of change is viewed with alarm. Huntington's proposal would be anathema to those charged with holding the alliance together. Third, what strategic objective would Huntington's strategy obtain? A successful invasion of East Europe would have NATO occupying Prague and most likely the Warsaw Pact would be in Frankfurt, astride the invasion forces line of communication. How such a situation could lead to favorable war termination possibilities for NATO is unclear. Finally, the proposal must assume that the Warsaw Pact nations will fight for the homeland and some plans must be made to contend with the twenty-five plus divisions of the Soviet strategic reserve. Neither of these contingencies are discussed adequately by proponents of a conventional retaliatory offensive.

Any of these objections would be enough to cast a shadow over the wisdom of pursuing a conventional retaliatory strategy, but the combination of the political and operational difficulties taken together should convince the prudent policymaker and strategist that the path to effective deterrence and defense leads elsewhere.

ENDNOTES

1. Christopher Layne, "Ending the Alliance," *Journal of Contemporary Studies,* vol. 6, no. 3 (Summer 1983), p. 8.

2. International Institute for Strategic Studies, *Strategic Survey, 1980-81* (London: International Institute for Strategic Studies, 1981), p. 37.

3. Samuel P. Huntington, ed., *The Strategic Imperative: New Policies for American Security* (Cambridge, Mass.: Ballinger Publishing Co., 1982), p. 3.

4. For just one example of the school of thought, see Jeffrey Record and Robert J. Hanks, *US Strategy at the Crossroads: Two Views* (Cambridge, Mass.: Institute for Foreign Policy Analysis, Inc., 1982).

5. Robert W. Komer, "Maritime Strategy Versus Coalition Defense," *Foreign Affairs,* vol. 60, no. 5 (Summer 1982), pp. 1124-1144; and Komer, "Is Conventional Defense of Europe Feasible?" *Naval War College Review,* vol. 35, no. 5 (September-October 1982), pp. 80-91.

6. Stephen S. Kaplan, ed., *Diplomacy of Power: Soviet Armed Forces as a Political Instrument* (Washington: The Brookings Institution, 1981), p. 668.

7. The original study was completed in 1979 under a contract for the Defense Advanced Research Projects Agency. While some revisions were made in light of the Afghanistan invasion, there were no substantive changes in the study's basic conclusions.

8. Dimitri K. Simes, "Deterrence and Coercion in Soviet Policy," *International Security,* vol. 5, no. 3 (Winter 1980/81), p. 86. Simes' research indicates that

> the majority of Soviet writers who claim that nuclear war is winnable . . . are neither top statesmen nor leading military personalities. As a rule, they are professors of Marxist-Leninist philosophy or of communist party history and are associated with the Military-Political Academy or with political indoctrination departments of other military educational institutions. This group of 'military commissars' plays an important role in what the Soviets call 'moral political preparation' for a possible war. It would be rather illogical for indoctrination experts to accept easily that there is no hope whatsoever in case of a major nuclear exchange. As spokesmen for the military establishment in general, these commissars may perform a function of warning about the sinister plans of 'imperialist forces' on the one hand, and on the other, improving the morale of the Soviet armed forces and population by telling them that even a nuclear war may be won by their country.

> Military commissars may be earnest in their optimist war-winning pronouncements, but their views for one reason or another are more and more overshadowed by a new orthodoxy . . . stressing the lack of political utility served from a nuclear war. (p. 88)

See also the work done by Robert Arnett, "Soviet Attitudes Towards Nuclear War: Do They Really Think They Can Win?" *The Journal of Strategic Studies,* vol. 2, no. 2 (September 1979), pp. 172-179; and David Holloway, *The Soviet Union and the Arms Race* (New Haven, Conn.: Yale University Press, 1983), pp. 29-64.

9. *Foreign Broadcast Information Service—Soviet Union* (June 16, 1982), p. AA1. Hereafter referred to as *FBIS-SOV*.

10. Lawrence T. Caldwell and Robert Legvold, "Reagan Through Soviet Eyes," *Foreign Policy,* no. 52 (Fall 1983), pp. 3-21.

11. *FBIS-SOV* (March 28, 1983), p. A3.

12. Francis J. West, Jr., "Conventional Forces Beyond NATO," in *National Security in the 1980's: From Weakness to Strength,* ed., W. Scott Thompson (San Francisco, Calif.: Institute for Contemporary Studies, 1980), pp. 319-336.

13. Henry Kissinger, *White House Years* (Boston, Mass.: Little, Brown, and Co., 1979), p. 182. Kissinger argued that in the National Security Council, there were three views on US-China relations: "One view (which we might call the 'Slavophile' position) argued that the Soviets were so suspicious of US-Chinese collusion that any effort to improve relations with China would make Soviet-American cooperation impossible. Those who held this view believed that we should give top priority to improving relations with the Soviet Union and, for this reason, should avoid efforts to increase contact with Peking. An opposing view (a kind of 'Realpolitik' approach) argued that the Soviets were more likely to be conciliatory if they feared that we would otherwise seek a rapprochement with Peking. This school of thought urged that we expand our contacts with China as a means of leverage against the Soviet Union. A third 'Sinophile' group argued that our relations with the Soviet Union should not be a major factor in shaping our China policy. Marginal actions to increase Soviet nervousness might be useful but fundamental changes in the US-China relationship should be guided by other considerations. Not surprisingly, I was on the side of the Realpolitikers."

14. "By showing the Soviets that we have the military capability and the national will to respond to aggression, we seek to deter such aggression in the first place. The determination and ability to move a credible American force rapidly and effectively changes the calculus for the Soviets; they must then consider the probability that any aggression by them will meet not only indigenous forces, but also those of the United States. Given such an ability on our part to meet them on the spot and our capability of shifting the geography of the conflict, the Soviets must consider the possibility that renewed aggression by them may lead to a much wider war, escalated both in intensity and geography." Harold Brown, *Annual Report of the Department of Defense, Fiscal Year 1982* (Washington: US Government Printing Office, 1981), p. 83.

15. See Francis J. "Bing" West, Jr., "NATO II: Common Boundaries for Common Interests," *Naval War College Review,* vol. 34, no. 1 (January-February 1981), pp. 59-67.

16. "For the region of the Persian Gulf, in particular, our strategy is based on the concept that the prospect of combat with the United States and other friendly forces, coupled with the prospect that we might carry war to other arenas, is the most effective deterrent to Soviet aggression." Caspar W. Weinberger, *Annual Report of the Department of Defense, Fiscal Year 1983* (Washington: US Government Printing Office, 1982), p. I-14.

17. The quotes are taken from Caspar W. Weinberger, *Annual Report of the Department of Defense, Fiscal Year 1984* (Washington: US Government Printing Office, 1983), p. 35. Nearly exact remarks were made by William P. Clark during a speech at Georgetown University Center for Strategic and International Studies in May 1982.

18. See Caspar W. Weinberger, *Annual Report of the Department of Defense, Fiscal Year 1985* (Washington: US Government Printing Office, 1984), pp. 37-38.

19. General Bernard W. Rogers, "The Atlantic Alliance: Prescriptions for a Difficult Decade," *Foreign Affairs,* vol. 60, no. 5 (Summer 1982), pp. 1145-1156.

20. International Institute for Strategic Studies, *The Military Balance, 1983-84* (London: International Institute for Strategic Studies), p. 16.

21. Computing the force requirements for a conventional retaliatory offensive is no easy task. The scenario that we used to determine manpower requirements was based on the following factors: First, the Huntington scenario as described in Chapter 2 was used for both NATO and Warsaw Pact forces. Second, NATO and Warsaw Pact force structures were based on current unclassified sources. Third, to avoid a static numerical comparison of the opposing forces, we used James F. Dunnegan's *NATO: Operational Combat in Europe in the 1970* (New York: Simulations Publications, Inc., 1973)—a commercial wargame—to game Huntington's proposals. Although several years old, the relative combat power ratios used in this game seemed accurate enough to estimate the approximate force requirements demanded by a retaliatory offensive. Fourth, it was assumed that tactical airpower for both sides was a neutral factor and that both sides had mobilized for about thirty days. Fifth, no additional Warsaw Pact forces were allocated after D-Day, but NATO was allowed to reinforce at will, regardless of manpower constraints. French units fought alongside NATO beginning at D-Day. Sixth, the USSR second strategic echelon, composed of twenty-four divisions from the Western Military Districts, was committed when the NATO offensive reached Leipzig and Prague.

When we ended the game on D + 21, Warsaw Pact forces were at the Weser-Lecht River line moving toward Frankfurt and NATO divisions were fighting for Prague and Leipzig. However, to insure that the NATO offensive was successful and that a Soviet breakthrough did not materialize in the north, thirty-one *new* NATO divisions were required during the course of the game: twenty US and eleven European. The United States had to commit eleven divisions just to deal with the Soviet second strategic echelon when Soviet forces counterattacked in the vicinity of Prague and Leipzig. The above does not pretend to be precise. However, even if the number of divisions needed was only one half of the requirement developed during the wargame, which does not seem implausible, this would mean that the United States would need an additional ten divisions in Europe on D-Day and another five allied divisions for a total of fifteen beyond the current program force. Although the wargame did not result in exact force requirements, it does give a dynamic characterization of the battle because it accounts for differences in training, mobility, firepower, and readiness among the various divisions of both sides. Moreover, it points to the need for a detailed evaluation of operational proposals such as Huntington's before they can be considered seriously. Finally, if it does nothing else, it highlights the important effect that the Soviet strategic second echelon can have which Huntington ignores in his proposal.

22. The American record in this regard is very poor. See William W. Kaufmann, *Planning Conventional Forces, 1950-1980* (Washington: The Brookings Institution, 1982).

23. *Ibid.,* pp. 23-24 and William W. Kaufmann, "The Defense Budget," *Setting National Priorities: The 1982 Budget* (Washington: The Brookings Institution, 1981), p. 152. POMCUS stands for prepositioning of materiel configured to unit sets.

24. Congressional Budget Office, *Rapid Deployment Forces: Policy and Budgetary Implications* (Washington: Congressional Budget Office, 1983), p. 25.

ABOUT THE CONTRIBUTORS

MALCOLM B. ARMSTRONG is a career fighter pilot with operational experience concentrated in Europe and NATO committed forces. He served four years as an F-111 pilot at RAF Upper Heyford, UK, during the initial deployment of that system to Europe. After graduation from Air Command and Staff College in 1974 and service in Korea, he served two years as Chief of the F-111 Operations and Training Branch at HQ Tactical Air Command. His command assignments include Assistant Deputy Commander for Maintenance, 58 Tactical Training Wing with the F-4, and Deputy Commander for Maintenance, 405 Tactical Training Wing with the F-15 and F-5, all at Luke AFB, Arizona, from 1979-1982. He is a graduate of the National War College Class of 1983 and is currently assigned as Vice Commander, 474 Tactical Fighter Wing with F-16 aircraft at Nellis AFB, Nevada.

VERNON V. ASPATURIAN is the Evan Pugh Professor of Political Science and Director, Slavic and Soviet Language and Area Center, Pennsylvania State University. He is the author of *Process and Power in Soviet Foreign Policy* and author, co-author, and co-editor of six other major works. He has been a consultant to RAND Corporation, the Army War College, Planning Research Corporation, the US Arms Control and Disarmament Agency, the Pacific Sierra Corporation, and HRB Singer. His articles on international politics, comparative foreign policy, and Soviet politics and foreign policy have appeared in numerous edited volumes and scholarly journals, including *American Political Science Review, Journal of Politics, Journal of International Affairs, Survey,* and *Problems of Communism.*

WESLEY K. CLARK is an Armor officer. His assignments have included tours with infantry and armored forces in the United States, Vietnam, and Germany. He was a White House Fellow serving as Special Assistant to the Director, Office of Management and Budget in the White House. He served two and one half years as Commander, 1st Battalion, 77th Armor, Fort Carson, Colorado. He is a graduate of the National War College Class of 1983, and is currently assigned as Chief, Army Studies Group, Headquarters, Department of the Army, at the Pentagon.

KEITH A. DUNN is the Senior Policy Analyst at the Strategic Studies Institute, US Army War College. He served in the US Army as a military intelligence officer. He is co-author of *Strategic Implications of the Continental-Maritime Strategy Debate*. His articles on US military strategy and the Soviet Union have appeared in numerous scholarly journals, including *Foreign Policy, Orbis, World Affairs, Soviet Union,* and *Naval War College Review.*

HOWELL M. ESTES, III is a career fighter pilot with extensive operational and staff experience in Europe. He served as an F-4 aircraft commander at Soesterberg AB, the Netherlands, in the early 1970s followed by a two and one half year tour at HQ US Air Forces Europe. After attending Air Command and Staff College in 1974, he served three and one half years on the Air Staff in the Europe/NATO Division, Directorate of Plans. He is a graduate of the National War College Class of 1983 and is currently assigned as the Deputy Director for Joint and NSC Matters, Directorate of Plans, Headquarters, US Air Force, at the Pentagon.

SAMUEL P. HUNTINGTON is the Eaton Professor of the Science of Government and Director of the Center for International Affairs, Harvard University. During 1977-78 he served at the White House as Coordinator of Security Planning for the National Security Council. He is the author, co-author, and editor of numerous books, including most recently *Living with Nuclear Weapons, The Strategic Imperative: New Policies for American Security,* and *American Politics: The Promise of Disharmony.*

CATHERINE McARDLE KELLEHER is a professor and head of the national security studies concentration at the School of Public Affairs of the University of Maryland. Educated at Mt. Holyoke College and at MIT, she currently holds a Ford International Security Affairs award for research on European security perspectives in the 1980s. Among her recent publications are *Sicherheit—Zum Welchem Preis?* (Munich: Olzog, 1983) as well as articles in the *American Political Science Review, Washington Quarterly,* and *International Security.*

JOHN R. LANDRY is a career Army officer with extensive command and staff experience in Europe. After graduation from the US Army Command and General Staff College in 1974, he served as an armored cavalry squadron executive officer for two years in Germany and was Special Assistant to the Supreme Allied Commander Europe, for another two years. He was an armored cavalry squadron commander in Germany followed by a tour as Chief of the Strategic Plans and Policy Division for Deputy Chief of Staff Military Operations and Plans, Headquarters, Department of the Army. He is a graduate of the National War College Class of 1983 and is currently assigned as Commander, 3rd Brigade, 4th Infantry Division (MX), Fort Carson, Colorado.

DANIEL S. PAPP is Professor of International Affairs and Director of the School of Social Sciences at Georgia Tech. He has served as Research Professor at the Strategic Studies Institute of the US Army War College and as Senior Research Associate at the Airpower Research Institute of the US Air University. His books include: *Vietnam: The View from Moscow, Peking, Washington; Communist Nations Military Assistance* (co-editor); *Contemporary International Relations: Frameworks for Understanding,* and *Soviet Policies Toward the Developing World During the 1980s* (forthcoming). His articles have appeared in a variety of journals, including *Soviet Union, Naval War College Review, Parameters, International Journal, Resources Policy,* and *Air University Review.*

KEITH B. PAYNE is Director of Research at the National Institute for Public Policy, Fairfax, Virginia. He received his Ph.D. from the University of Southern California and was on the Senior Professional Staff of Hudson Institute from 1979-82. Dr. Payne is the author of *Nuclear Deterrence in US-Soviet Relations,* co-author of *Nuclear Strategy: Flexibility and Stability,* and editor and contributor to *Laser Weapons in Space: Policy and Doctrine,* and *Playing with Fire: The Implications of a Nuclear Freeze.* His articles have appeared in various journals, including *Foreign Policy, Orbis, The Washington Quarterly, Comparative Strategy, The Wall Street Journal, Air Force Magazine,* and *USA Today.*

RICHARD HART SINNREICH, Lieutenant Colonel, US Army, is the Army Fellow at Georgetown University's Center for Strategic and International Studies. He has commanded artillery units in Germany, Vietnam, and Korea, and has served on the Army, Joint, SHAPE, and NSC Staffs. His previous publications include chapters in several books on foreign and defense policy, and articles in *Orbis, Army,* and the *Field Artillery Journal.*

WILLIAM O. STAUDENMAIER is a Colonel in the US Army and is the Director of Strategy for the Center for Land Warfare, US Army War College. He is a graduate of the University of Chattanooga and Pennsylvania State University. Colonel Staudenmaier served in combat in Vietnam as a District Advisor and in various staff assignments at the Department of the Army. He has contributed chapters to several books dealing with military strategy and defense policy. His articles have appeared in *Foreign Policy, Orbis, Naval War College Review, Military Review, Army,* and *Parameters.* He is co-author of the forthcoming *Strategic Implications of the Continental-Maritime Strategy Debate.*

BOYD D. SUTTON has been a specialist in Soviet and European military subjects for most of his career. Following Army assignments in Germany, Vietnam, and at the Defense Intelligence Agency, he joined the CIA and has analyzed Soviet and Warsaw Pact command and control, strategic and theater nuclear forces and most recently was Chief of the Soviet/Warsaw Pact Ground Forces Branch. Mr. Sutton is a graduate of the National War College Class of 1983 and is currently assigned on rotation from CIA to the Office of the Deputy Assistant Secretary of Defense for European and NATO policy.

ADDITIONAL CONFERENCE PARTICIPANTS

COL Keith A. Barlow, Director, Strategic Studies Institute, US Army War College.

Dr. Richard K. Betts, The Brookings Institution.

Dr. John Despres, Director, Strategic Concepts Development Center, National Defense University.

Mr. Louis Finch, Deputy Director, Foreign Affairs and National Defense Division, Congressional Research Service.

LTC John Fulton, US Army Training and Doctrine Command.

LTG Richard Lawrence, President, National Defense University.

CAPT Richard Life, Strategic Support Team, Commander-In-Chief, Atlantic Fleet.

Dr. Stephanie Neuman, Columbia University.

MG William E. Odom, Assistant Chief of Staff for Intelligence, Headquarters, Department of the Army.

Dr. Alan N. Sabrosky, Director of Studies, Strategic Studies Institute.

Mr. Walter Slocombe, Caplin and Drysdale.

BG Anthony A. Smith, Office of the Assistant Secretary of Defense (International Security Policy).

Dr. Shirin Tahir-Kheli, Policy and Planning Staff, Department of State.

Dr. Edward L. Warner, III, RAND Corporation.

Mr. Bruce Weinrod, Heritage Foundation.

Dr. Samuel F. Wells, Woodrow Wilson Center.

GEN John A. Wickham, Jr., Chief of Staff of the US Army.

INDEX